Alternative Concrete – Geopolymer Concrete

Emerging Research and Opportunities

Adrian LĂZĂRESCU

Henriette SZILÁGYI

Cornelia BAERĂ

Andreea HEGYI

Published by **Materials Research Forum LLC**
Millersville, PA 17551, USA

Published as part of the book series
Materials Research Foundations
Volume 109 (2021)
ISSN 2471-8890 (Print)
ISSN 2471-8904 (Online)

Print ISBN 978-1-64490-152-6
ePDF ISBN 978-1-64490-153-3

Distributed worldwide by

Materials Research Forum LLC
105 Springdale Lane
Millersville, PA 17551
USA
http://www.mrforum.com

Printed in the United States of America
10 9 8 7 6 5 4 3 2 1

Table of Contents

Introduction

Concrete is one of the most versatile, durable and reliable building material, also being the most used one, which also involves large amounts of Portland cement as main, conventional binder for concrete production. Worldwide, the use of concrete is considered to be overcome only by the overall use of water. Environmental problems associated with the production of Portland cement are very well known and they represent a closely monitored problem in terms of carbon dioxide amount released into the atmosphere during its production, by lime stone calcining using fossil fuels for burning. Each ton of produced cement requires 60, up to 130 kgs of fuel oil or its equivalent, depending on the type of cement and the manufacturing process being used, and about 110 KWh of electric generated power [1]. A ton of manufactured OPC releases between 0,8 and 1,1 tons of CO_2 into the atmosphere, as a consequence of burning fuel and calcining limestone [2]. In 2014, the global cement production was about 4.2 billion tons [3], indicating an increase compared to the previous years.

The industry is developing at an accelerated pace, and accordingly, the use of modern building materials in the economic world is more than necessary. The large amount of fly ash, waste or by-product associated to energetic industry, can create new opportunities for using it as OPC substitute in manufacturing an alternative concrete. When used together with Portland cement in the production of concrete, as partial replacement of OPC, fly ash reacts with calcium hydroxide during the hydration process, in the presence of water. Nowadays, the use on a larger scale of materials derived from industrial by-products in the manufacture of concrete has substantially increased with the development and the corresponding demand, at a global rate, of producing special concrete: self-compacting concrete, high-strength concrete, high performance concrete, etc. [4].

New binding materials, called "geopolymers" were introduced in 1987 by Davidovits to describe a group of mineral binders, with a chemical composition similar to the natural zeolite materials, but with an amorphous microstructure. He described that "binders could be produced by a polymeric reaction of alkali liquids with silicate and aluminum from source materials of geological origin or with residual materials such as fly ash" [5]. Geopolymer cements are currently under development, with research mainly encouraged by the need to reduce CO_2 emissions globally. Having excellent mechanical properties and resistance in aggressive environments, these materials represent an opportunity for both the surrounding environment and for engineering, an alternative to traditional technology [6].

The concept of Alkali-activated Geopolymer Materials (AAGM) as an alternative to Portland cement has been known since the 1980s. Also, the durability of the materials obtained through this process has been demonstrated over the years in Belgium, Finland, Russia, China and recently, Australia and Great Britain. At the beginning of the 1990s, most of the research in the field of alkali-activated geopolymer materials has been focused on the microstructure of these materials and much less on the studies of the predictions concerning the properties and engineering characteristics of the materials, the load capacity in different types of conditions, strength, etc., and of the possibilities of their assessment. Until recently, the field of alkali-activated materials was looked upon more as an academic curiosity, with potential niche applications, but never as a substitute for Portland cement in the construction industry.

There are some good reasons why alkali-activated materials have not received the same attention in the building materials industry as concrete that uses Portland cement does. First of all, limestone is present in abundance all over the world, which makes it easier to produce Portland cement, being a very affordable material. In contrast, the necessary materials for producing alkali-activated materials (fly ash, slag, etc.) are not available everywhere. The second, and perhaps one of the most important reasons why research on alkali-activated materials is less discussed, is that Portland cement and concrete have a history of more than 150 years in the construction industry, while applications of alkali-activated materials are limited and isolated, with few notable examples. Furthermore, all the regulations regarding the building materials production (testing methodologies and performance, etc.) show a clear dependence to the OPC as main binder in their composition, even if sometimes other materials are regarded as binding agents to produce concrete.

While geopolymer technology is not currently self-sufficient to satisfy the economic and ecological drivers, environmental requirements in terms of reducing carbon dioxide emissions have become a sufficient reason for the construction industry to take into account producing this type of material.

Worldwide, research on alkali-activated geopolymer materials exists, but their actual production differs greatly, due to variation of factors influencing this type of material. Unlike concrete made with ordinary Portland cement, geopolymer concrete does not form calcium silicate hydrates for its matrix formation and for obtaining mechanical properties. In the case of geopolymer materials, silicon (Si) and aluminum (Al) react with an alkaline solution, forming an aluminum-silicate gel that unites the binder with the aggregates and provides the necessary strength of the material. Thus, the source material and the alkaline activator are the two major constituents of the geopolymers, the strength of this type of material is dependent on the direct nature of the materials used and the type of alkaline liquid.

For the production of geopolymer concrete, choosing the source materials depends largely on their availability and cost, the type of process to which they are subjected and specific requests from users. There is a lot of research regarding the production of alkali-activated fly ash-based geopolymer materials however, their implementation requires special procedures, because there are not yet unitary regulations governing the manufacture of this type of material. They are carried out on the basis of special technical specifications, complementary to existing norms.

Both circular economy and sustainable development are concepts that focus on resource efficiency, which implies a complex waste management process with a high degree of recycling and recovery. Designing resilient, sustainable buildings, using recycled materials that can replace some or all of the cement, leads to the development of a sustainable, low-carbon economy. At the moment, Romania, it is not ready for the implementation of these concepts, given the modest effective use of these by-products and also in conjunction with a significant increase of the natural resources consumption of for the production of electrical energy in a conventional manner.

This upward trend, inversely proportional to the strategy of EU countries and successfully implemented in some, can be countered at the industrial level by solutions for the massive reintroduction of industrial waste into the active production circuit of materials used in construction. Innovative materials such as alkali-activated geopolymers can become a viable alternative for integrating concrete structures into the eco-intelligent structures category.

Since the 90s, research in the field of alkali-activated materials has increased dramatically across the globe, most of them focusing on the development of materials with corresponding performance when related to the conventional concrete, made with local materials, available in the respective area. Any new and unconventional technology is difficult to translate into ordinary practice, since the existing standards for concrete do not include the specific directions for geopolymer materials production. Current standards regulate products that are already accepted on the market, ensuring that they comply with certain rules and have certain characteristics and qualities.

The large amount of fly ash produced in Romania can create and encourage the development of new opportunities for its use as total substitute of the Portland cement for the production of new types of concrete and mortars in the construction industry.

The general aim of this research is to make significant contributions in understanding and deciphering the mechanisms of the realization of the alkali-activated fly ash-based geopolymer materials and, at the same time, to present the main characteristics of the materials, components, as well as the influence that they have on the performance of the mechanical properties of the material.

This research is fundamentally motivated by the need, identified both at global and national level, to implement the principles of sustainable development, sustainable consumption of natural resources, recovery of existing waste and prevention of generating new ones, in the global ecological context of drastic reduction of harmful effects related to pollution, destabilization of ecosystems, global warming and all other related elements. Therefore, the present research deals the in-depth research of the potential recovery of the fly ash and using it as a raw material for the development of new construction materials, offering sustainable solutions to the construction industry. The way to achieve this target is represented by producing fly ash-based alkali-activated materials using Romanian local raw materials.

The research also contributes to the opportunity and need of continuous and updated studies on the identification and capitalization of various materials resulting from technological processes in the energy industry, metallurgical industry, etc. In addition to the economic advantages, another important aspect is the environmental aspect; the re-use of waste materials and industrial by-products (source of environmental pollution) in the economy, accounting in addition to the alternative use of the raw material, a beneficial process for "greening" the area. Also, the total replacement of cement in producing innovative cementitious composites contributes to the reduction of carbon dioxide emissions.

It is clear that the obstacles that alkali-activated geopolymer materials face in their use are high, but substantial progress is being made in terms of their development and marketing in recent years.

The increased interest in such materials is mainly due to the needs in the reduction of carbon dioxide emissions. The carbon footprint of this type of material is much lower than that of concrete using ordinary Portland cement. It can be said that it is a conscious niche rather than one generated by the present or future carbon dioxide emissions legislation.

Acknowledgement

This book is based on research conducted within the framework of a PhD thesis and had the purpose to obtain and characterize alkali-activated fly ash-based geopolymer materials, obtained using local raw materials from Romania and to study their mechanical properties. All the research and data collection were conducted within a complex interdisciplinary system between the Technical University of Cluj-Napoca and NIRD URBAN-INCERC (National Institute for Research and Development in Urban Planning and Sustainable Territorial Development URBAN-INCERC) Cluj-Napoca Branch, Romania.

Also, the authors would like to thank Prof. Michael GRANTHAM of Sandberg LLP (former Visiting Professor at the University of Leeds and Queen's University Belfast and a past President of the Institute of Concrete Technology, U.K.), Prof. PhD. Eng. Călin Mircea (Technical University of Cluj-Napoca) and Assoc. Prof. PhD. Eng. Andrei Victor SANDU (Faculty of Materials Science and Engineering, Gheorghe Asachi Technical University of Iasi, Romania and President of Romanian Inventors Forum) for their support and contribution in technical editing, language editing and proofreading of this book manuscript.

Materials Research Forum LLC
https://doi.org/10.21741/9781644901533

Chapter 1

Global Perspectives on the Production of Industrial By-Products in the Context of "Circular Economy"

Concerns about CO_2 emissions are the creation of trade and regulatory incentives to reduce coal emissions in many countries around the world. Despite this pressure, the share of coal in the world electricity production remains above 38%, with coal consumption increasing in recent years after a short period of annual decline [8]. This recent increase in coal consumption is largely driven by economic growth in large developing countries such as China, India and parts of Southeast Asia.

In many developed economies, a reduction in coal-fired electricity production is under way, with a total elimination of coal power over the coming decades. The transition from the use of coal for electricity generation, is accompanied in several countries by an increase in the use of fossil fuels, such as natural gas, alternative fuels (such as nuclear energy, biomass), and increased use of renewable energy, including wind, solar, hydro, and geothermal. The choices of energy production and speed of production to transition to these alternatives is highly dependent on the political, economic, and geographical conditions of the areas.

The production volume of coal combustion products (CCP – Coal Combustion Products) is directly correlated with the burning of coal in thermal power plants. Their commercial and environmental value is well established and directly proportional to their use in in the manufacture of building materials. The management and use of CCPs is similar in most countries, policy-makers, encouraging producers and buyers to increase their usage in various applications.

In order to make more efficient use of industrial by-products, producers, and policy makers need to understand and comply with the regulatory conditions, market demand, product quality, etc. In some countries, most CCPs are already used in accordance for the production of building materials in accordance with existing standards. This occurred due to a "mature" market, with a constant demand in quality products used in building materials. On the other hand, the lack of adequate standards, poor "education" on the market and regulatory barriers - such as the designation of CCPs as waste and not as resources or raw materials - has led to reduced used rates of large volumes of stored CCPs.

Since the global amount of CCPs is directly related to energy production through the use of coal, it is necessary to understand the current state and prospects for global coal consumption and future demand.

Currently, around 7.700 Mt of coal is used worldwide by a large number of sectors, including electricity generation, but also other industries, such as iron and steel production, cement manufacture etc. [9].

Growth in coal demand in the near future is concentrated in India, Southeast Asia and several other Asian countries. Demand for coal is expected to decline in Europe, Canada, the United States and China. As a result of these contrasting trends, global demand for coal will show only slight growth over the next decade. Although coal-fired power generation will increase in absolute terms, demand is still expected to decline due to increased use of renewable energy [2].

In order to ensure the sustainability of both natural and economic ecosystems, it is necessary to mitigate the negative impact of industrial activities on the environment. Carbon dioxide emissions from industry have been the subject of national legislation for many years. The requirements for power plants have led to actions which they have been obliged to comply with. This was a successful example of the adoption of pollution reduction technology, with additional environmental benefits accumulated through the role of efficient use of CCPs resulting from electricity production.

European Union Directive regarding industrial emissions (European Union, 2010) defines BAT as „the most effective way in the development of the activities and the way of operation in the industry, indicating the appropriateness of the practice of certain techniques for delivery of new technologies to reduce the emission levels of pollutants and others draw the conditions designed to prevent and, where that is not practicable, to reduce as the much of them, therefore, reducing the negative impact on the environment".

Since the use of industrial by-products is either standardized or linked to a special project, national regulations for their use in different applications should be considered. The easiest way to serve existing markets is through product standards. Therefore, a compilation of standards for the use of fly ash in the production of alkali-activated geopolymer materials requires increased attention, which can lead to high demand for the production of this type of material.

Global production of industrial by-products produced by burning coal is more than 1.2 billion tons annually, almost doubled in the last 5 years. Usage rates vary widely due to different regulatory environments, market education and market conditions.

Materials Research Forum LLC
https://doi.org/10.21741/9781644901533

The large amount of fly ash from power plants in Romania, can lead to new opportunities for the use of this residue as a substitute for Portland cement for the manufacture of new kinds of materials, in particular alkali-activated fly ash-based geopolymer concrete.

Chapter 2

Retrospective Regarding Research in the Field of Alkali-Activated Geopolymer Materials

2.1. Geopolymers and geopolymerisation – chemical characteristics

Because for the production of new materials, most of the time the rules or standards in force do not apply, it is necessary to understand the chemical phenomena that occur and how they lead to the creation of a new type of alternative and innovative material. The development of Geopolymer Cement and Geopolymer Concrete opens up new opportunities for the production of environmentally friendly cement and concrete. Geopolymer is a cementitious aluminum-silicate, amorphous material, which can be synthesized by the polycondensation reaction between a geopolymer material (fly ash) and alkali polysilicates. This process is called geopolymerization [10]. This innovative technology allows several aluminum-silicate materials to be turned into products called geopolymers or inorganic-polymers.

In a simplified method, it can be said that geopolymers can be synthesized by alkaline activation of materials that are rich in SiO_2 and $Al2O_3$ [5,6,11]. The entire process is still subject to a lack of data consistency. The context of the mechanisms of the geopolymerization shall cause the dissolution of Al in an alkaline medium, transportation (orientation) of the solute, forming a 3D aluminum-silicate network structure [12].

More than 65% of the Earth's crust consists of Si-Al compounds and that is why it is important to study new types of materials and to understand how they can contribute to the development of new types of concrete, that also have good mechanical properties.

In 1999, Palomo proposed that pozzolanic materials (blast furnace slag, fly ash) can be activated "using alkaline liquids, to form a binder and completely replace the use of Portland cement in the production of concrete" [12], but only in 2001, research on geopolymer concrete began to take shape. There are several specialized publications dealing with this topic, but the literature is quite limited in terms of technology of producing this type of material, many of them regarding only small research and not full-scale elements. Mehta suggested in 2002 using fewer natural resources, using less energy and minimizing carbon dioxide emissions over long periods to produce new green concrete [13]. He demonstrated that a reduction in carbon dioxide emissions can be achieved by decreasing the amount of calcined material, by decreasing the amount of cement in concrete and by decreasing the number of buildings made with conventional cement. Although this is, at least in the near future, too difficult to achieve, many efforts are made to reduce the use of Portland cement in the manufacture of concrete. For the production of alkali-activated geopolymer concrete, the choice of source materials largely depends on their availability and cost, the type of process to which they are subject and specific requests from users.

The empirical formula for the entire geopolymerization process is presented in Eq.1 [14]:

$$Mn\{-(SiO_2)z-AlO_2\}n,wH_2O \tag{1}$$

Where: M – is the alkaline element, which can be:

- K (potassium);

- Na (sodium);

- Ca (calcium);

The symbol "-"indicates the presence of bonding between elements;

n – is the degree of polymerization;

z – is the Si/Al ratio, which can be 1, 2, 3 …. to 32;

When the two components of the geopolymer material react (the reactive solids and the alkaline activator) an aluminum-silicate network is formed, creating a hard, water resistant material [13]. The geopolymerization reaction can also be chemically expressed according to Fig. 2.1.

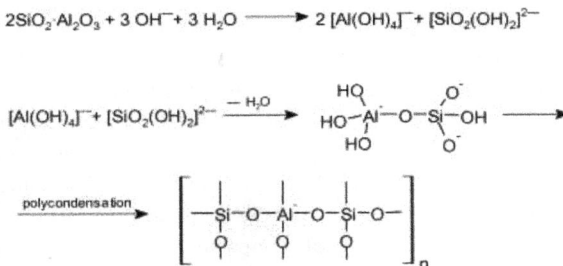

Figure 2.1.   Geopolymerization reaction.

In the case of alkali-activated fly ash-based geopolymer materials, the dissolution process (geopolymerization reaction) of Si and Al occurs when the fly ash comes into contact with the alkaline solution. Then the larger molecules will condense and form a gel that, under the alkaline attack occurring on the surface of the particles, will lead to an expansion of the formed gel covering the remaining voids to form a matrix. This alkaline attack occurs both from the outside and from the inside of the molecules. As a result, the reaction product is generated both inside and outside the spheres until the ash particles are completely dissolved (Figure 2.2) [15].

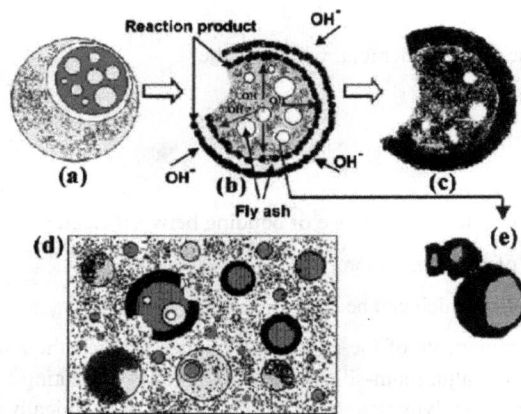

Figure 2.2.   Descriptive model of alkali-activation of fly ash.

The reaction of those materials in a highly alkaline medium result in the breaking of the Si-O-Si bonds, subsequently resulting in new phases and the mechanism of its formation appears to be a process that includes the presence of a solution (synthesis via solution). The most important part of this reaction is the penetration of Al atoms into the original Si-O-Si structure, when, more often aluminum-silicate gels are formed. The composition of these gels can be characterized by Eq. 1. C-S-H and C-A-H phases may also have their origin in their direct dependence on the chemical composition of the materials used and the conditions for producing the reactions [10]. Also, the concentration of solid matter plays an important role in the alkaline activation process [16].

The chemical reactions that produce alkali-activated fly ash-based geopolymers may be suspected in the next discussion [17]:

- hydrating process for vitreous silica, with pH>12 is:

$$SiO_2 + 2OH^- <> SiO_3^- + H_2O \tag{2.2}$$

- and Al_2O_3 is hydrated according to:

$$Al_2O_3 + 2OH^- <> 2AlO_2^- + H_2O \tag{2.3}$$

- and CaO and MgO react as following:

$$CaO + H_2O <> Ca^{2+} + 2OH^- \tag{2.4}$$

- with Na_2O and K_2O reacting:

$$Na_2O + H_2O <> 2Na^+ + 2OH^- \tag{2.5}$$

- Fe_2O_3 reacts according to:

$$Fe_2O_3 + 3H_2O <> 2Fe^{3+} + 6OH^- \tag{2.6}$$

- and TiO_2 hydrates:

$$TiO_2 + OH^- <> HTiO_3 \tag{2.7}$$

The chemical reaction of fly ash is:

$$SiO_2.\alpha Al_2O_3.\beta CaO.\gamma Na_2O.\delta Fe_2O_3.\varepsilon TiO_2 + (\beta+\gamma+3\delta)H_2O + (2+2\alpha+\varepsilon)OH^- <> SiO_3^{2-} + 2\alpha AlO_2^- + \beta Ca^{2+} + 2\gamma Na^+ + 2\delta Fe^{3+} + \varepsilon HTiO^{3-} + (1+\alpha)H_2O + 2(\beta+\gamma+3\delta) \tag{2.8}$$

From the reactions mentioned above, the thermodynamic properties (enthalpy and entropy) of the fly ash can be seen [17]. It can be seen that silicon, aluminum oxide and titanium oxide consume hydroxides, and CaO and MgO (earth alkalis), Na_2O and K_2O (alkaline substances) and iron oxides produce hydroxides [11]. Fly ash contains high percentages of aluminum and amorphous silicon, which makes it usable for the production of geopolymers. The only important factor in the geopolymerization process is the type and concentration of the alkaline activator, which can affect the dissolution of the fly ash.

2.2. Geopolymers and geopolymerization – mechanical characteristics

For a material used in the construction industry, mechanical behavior is a basic property, which makes it optimal for a specific application. Since geopolymer materials are a novelty in this area, compressive strength is an important factor. Since 1950, good compressive strength, workability and durability of these new materials have been perceived as better than those of ordinary Portland cement concrete. However, the compression behavior of geopolymers varies depending on the used materials and the production methods they are subjected to [18]. In order to obtain a geopolymer with a good compressive strength, one should take into account the type and the molar ratios of the oxides in the source material, the pH of the alkaline solution, and the solubility of the source material in alkaline activator [19].

In 1994, Davidovits introduced three "key parameters" to produce geopolymers with good mechanical performances. For this to be possible, the following conditions must be met [5]:

$$0.2 < Na_2O/SiO_2 < 0.28$$

$$3.5 < SiO_2/Al2O_3 < 4.5$$

$$15 < H_2O/Na_2O < 17.5$$

The mechanical properties of the geopolymer materials are directly affected by the way in which the rich Al-Si materials dissolve in the alkaline activator and their microstructural reorganization when the reaction occurs. It has been shown that flexural strength, compressive strength and apparent density of concrete increased with the increase of the NaOH solution concentration [20], which also increased the amorphous content of the product [21].

Although the dissolution of the source material increases with the increase of the concentration of the alkaline solution, a larger amount of NaOH or KOH in the aqueous phase leads to a decrease in the SiO_2/Na_2O ratio, affecting the reaction [22]. Using alkaline solution composed of alkaline hydroxide and dissolved silicate is beneficial for the compressive strength of the material. Dissolved silicon not only optimizes the SiO_2/Al_2O_3 and Na_2O/SiO_2 ratios in the mixture, but also catalyzes the polycondensation phenomenon by $SiO4-$ monomers, which initiates the polymerization phenomenon between AlO_4^- and SiO_4^- [18]. Therefore, high compressive strength can be obtained using an activator composed of soluble silicate and alkaline hydroxide.

It should be also noted that there is a limit for the addition of silicates in the mixture (Si/Al = 1.90), and that high Si/Al ratios are not recommended as they have a negative effect on the mechanical properties [23]. A high Si/Al ratio increases the porosity of the material by increasing the percentage of undissolved material, decreasing the compressive strength of the geopolymer material.

2.3. Types of alkaline activators used in the synthesis of geopolymers

In geopolymer cement, aluminum-silicate minerals have a major component of SiO_2 (silicon dioxide) and Al_2O_3 (aluminum oxide) which are dissolved from the raw material (fly ash) when in contact with a highly alkaline solution. There are a lot of alkaline activators which have been used in the synthesis of the geopolymers, for example potassium/sodium hydroxide, (KOH/NaOH), potassium/sodium silicate (K_2SiO_3/Na_2SiO_3), sodium carbonate (Na_2CO_3), calcium hydroxide ($Ca(OH)_2$), or a combination of these solutions [24,25].

However, the most used alkaline activators in the synthesis of geopolymers can be synthesized as follows:

Sodium hydroxide solution (NaOH)

NaOH solution is normally used to produce geopolymer cement due to its widespread availability and is cheaper than other alkaline solutions [26]. Its main role is to provide an alkaline medium of hydroxide ions (OH^-) for the dissolution of the source material.

Sodium silicate solution (Na_2SiO_3)

Na_2SiO_3 solution or water-glass is normally used in geopolymer synthesis as an alkaline activator and another source of silicon dioxide (SiO_2). It is also cheaper than potassium silicate solution (K_2SiO_3) when produced in large quantities [27]. Like other alkaline activators, the mechanical properties of geopolymers increases with an increase in the concentration of this solution. However, by using only the sodium silicate solution may not achieve the same levels of mechanical resistance of the geopolymer material, due to the fact that the rate of dissolution of Si and Al in the source material is lower than that of the use of the OH-compound [28].

Sodium hydroxide and sodium silicate solution (NaOH and Na_2SiO_3)

A combination between sodium hydroxide solution and sodium silicate solution is one of the most used alkaline activator for the synthesis of geopolymers. It is known that soluble hydroxide dissolves aluminum minerals from source materials, while soluble silicon improves the poly-condensation of the geopolymers and also controls the amount of silicates in the mixtures. The optimal proportion of Na_2SiO_3 and NaOH is therefore an important factor in the synthesis of geopolymers [25-28].

2.4. Alkali-activated geopolymer materials mix-design

2.4.1. Properties of the alkaline activator

It has been shown that the compressive strength of the alkali-activated geopolymer materials increases, in general, with the increase in the concentration of the specific alkaline activators [11,16,29]. A higher concentration of the NaOH solution may produce stronger Si-O-Si bonds and improve the dissolution of source materials in the presence of the activators [30]. The optimal alkaline concentration also varies by a large number of conditions and factors that must be taken into account in the mix-design:

Alkaline activator to source material ratio

This parameter is often used in the design of alkali-activated geopolymer mixtures in order to define the alkaline dosage and at the same time the water content of the designed mixture. In most cases, the source material used for the production of AAGM is fly ash, in which case the notion of the ratio between alkaline liquid and fly ash (AA/FA) is introduced. The effect of this parameter on the compressive strength of the alkali-activated geopolymer concrete has been intensively studied [26,31] and the recommendation for obtaining both good mechanical properties and satisfactory workability is that it varies between 0.35 and 0.5 [32,33]. However, these ratios may only apply to certain properties of the source materials used for the production of the geopolymer material.

Moreover, since water in the mixture plays a very important role in the hardening process, the ratio of water and solids should be closely monitored. The total amount of water in the mixture is given by the sum of the amount of water in the solutions of the alkaline activator and any extra water that is added (w/s). The total amount of geopolymer solids is represented by the amount of fly ash used, the amount of aggregates, as well as the percentage of solids in the constituents of the alkaline activator. Literature proposal for this ratio is between 0.18 and 0.36 [26,34,35].

However, other studies have shown that these ratios cannot be fully applied due to the properties of the raw material used (chemical composition, fineness, specific surface, etc.) [36-39].

Na_2SiO_3 to NaOH solution ratio

The optimal dosage between the solutions of the components of the alkaline activator is an important factor in the synthesis of geopolymer materials. Studies have shown that for a source material with a low calcium content, the geopolymer material achieves satisfactory performance when this ratio is 1.0/1.5 or 2.0 [40].

2.4.2. Heat curing procedures

As for the geopolymerization reaction to take place as quickly as possible and to obtain alkali-activated materials with high mechanical properties and high durability it is necessary that the hardening temperature of the mixtures to be above the normal, room temperature, thus a heat treatment procedure is necessary. The heat treatment procedure can be performed at different temperatures and different periods of time, this becoming another important factor in determining the final performance of the geopolymer material [41,42].

The *effect of the curing temperature* on the alkali-activated geopolymer materials has been demonstrated by numerous studies which have shown that for the treatment temperatures between 30 and 90°C, as the treatment temperature increased, improvements were observed in the geopolymerization reaction and also increases in the mechanical properties of the material, at very young ages [43]. An increase in the treatment temperature over this interval showed that it led to a decrease in the performance of the obtained geopolymer material [44]. The optimum temperature for obtaining alkali-activated geopolymer materials with high performances is between 60 and 75°C [39,45,46].

It can be said that the curing temperature is one of the most important factors affecting the mechanical properties of alkali-activated geopolymer materials. Although above ambient temperatures it produces materials with high strength, too high temperature can lead to contractions and even cracks in the process of hardening of the material and also leads to a decrease in their mechanical performance.

The *effect of the duration of heat treatment* can also affect the mechanical properties of alkali-activated geopolymer materials. For the most widely used treatment temperatures (between 40 and 60°C), the duration of heat treatment varies between 4 and 96 hours, depending on the design principles adopted. However, periods of 6, 12, 24, 48 and 72 hours are most often used, being constant parameters for finding and studying other influences on the performance of the material [47-49].

Studies have shown that a treatment duration between 12 and 24 hours is optimum for the development of the mechanical properties of the geopolymer materials and also in terms of economic approach, although a longer treatment period could lead to higher mechanical strength [13,50,51].

For the heat curing regimes, there are a lot of possibilities both for their duration and temperature in order to achieve the desired mechanical properties of the geopolymer material. As mentioned above, the optimal treatment conditions depend mainly on the properties of the raw material used, alkaline activators, water content etc. [49]. Heat treatment directly affects not only the degree of geopolymerization, but also the water evaporated in the structure of the geopolymer matrix [52]. When water is rapidly released due to high temperatures or prolonged treatment periods, micropores may shrink, damaging the mechanical properties of the material [50]. Based on the above factors, the optimum temperature and hardening period for the production of alkali-activated geopolymers are between 40 and 70°C, respectively 8 to 24 hours.

2.5. Properties of alkali-activated geopolymer materials

There is a huge volume of specialized literature that treats the properties of alkali-activated geopolymer materials, which meet the testing standards of OPC. Alkali-activated geopolymer materials exhibit good physical and chemical properties. Their main properties, obtained by appropriate procedures and conditions, can be summarized as follows:

2.5.1. Mechanical properties of alkali-activated geopolymer materials

Setting time and the development of high mechanical strength at very young ages are among the most important features in the construction industry. Setting of ordinary Portland cement concrete is a time-dependent chemical process. In order to obtain a geopolymer material in a short time it involves heat treatment and at the same time the use of materials with high properties for its production [26,44].

Compressive strength is the main parameter according to which the mechanical characteristics of alkali-activated geopolymer materials are evaluated, being a simple parameter to determine [53]. Alkali-activated geopolymer concrete can reach strengths of up to 95 MPa at early ages (7 days) [54], which can be considered the equivalent of ultra-high strength concrete. Moreover, other tests performed on geopolymer concrete revealed that this type of material may have similar or even better properties than ordinary Portland cement concrete, for example: flexural strength [24], tensile splitting [55], etc. These mechanical properties are achieved only with a complex geopolymerization process, the materials used, their properties, and the specific use.

Shrinkage is characterized by the decrease in concrete volume over time and is a phenomenon independent of external factors, which leads to cracks in the material and also to losses in mechanical performance. Studies have shown that alkali-activated geopolymer materials had superior performance in terms of this parameter [54,56] compared to ordinary Portland cement concrete. Also, geopolymer materials have a low water absorption due to the dense matrix of the binder [57], largely determined by the amount of water used to produce the material, which determines its density [58].

2.5.2. Durability of alkali-activated geopolymer materials

Concrete or cement can be damaged by freeze-thaw actions, when their pores are filled with water. Studies have shown that alkali-activated geopolymers exhibit good freeze-thaw resistance by using alternating cycles between -15 and +60°C for four months. It was found that the analyzed material did not show any degradation following the application of the cycles. However, in terms of their mechanical strength, this was represented by a slight decrease [47].

Corrosion is also a destructive attack both by chemical and electrochemical reactions on reinforced concrete elements. Chemical attacks that produce physical damage are called corrosion-erosion or corrosive wear actions [59]. The worst environments for reinforced concrete elements, which generate corrosion phenomenon, are marine ones or are due to events involving high acidity. Studies have shown that the properties of the geopolymer materials are better than those of using Portland cement in the mixture, due to the more homogenous structure of the binder [25,46].

2.5.3. Remarks

The properties of alkali-activated geopolymer materials depend mainly on the important factors that could affect the development of this type of material and include: the characteristics of constituent materials, alkaline activators, heat treatment regimes, etc. It is obvious that due to the multitude of factors that could influence the geopolymerization reaction, the specificity and conditions met for each proposed mixture generate unique characteristics. However, in most cases, studies have shown that the properties of geopolymer materials were similar or even better than those of ordinary Portland cement concrete, when undergoing similar tests [55,60].

Already implemented in some countries, alkali-activated geopolymer concrete can be a great alternative to ordinary Portland cement concrete, the production of such materials, sensitive from the chemical point of view, and at the same time the need for heat treatment in order to obtain certain superior mechanical properties, are the current limitation of this type of material.

2.6. Factors influencing alkali-activated geopolymer materials

2.6.1. Physical and chemical properties of fly ash

The shape and size of the fly ash particles (fineness) directly affect the mechanical properties of geopolymer materials after alkaline activation. Small particles, with high specific surface area, with bigger fineness, increase the reaction level during the geopolymerization process. The dissolution rate, transport of ions, the formation of chemical bonds, etc., control the setting time and the final stage of the formation of the geopolymer [61,62]. The compressive strength was also higher due to the change in the structure of the geopolymer binder, with a higher rate of dissolution of the particles of fly ash into the alkaline environment [63,64]. Various tests in which raw fly ash and processed fly ash were used in order to evaluate the behavior of the geopolymerization process and also the final mechanical strength of the material [36-39,65,66].

Also, the chemical properties of the source material (fly ash) determine the properties of geopolymers, especially the amount of calcium oxide (CaO), but also the water/fly ash ratio [12]. The molar oxide ratio present in the mixture is also an important factor in the geopolymerization process. It is given by the percentage of calcium oxide (CaO), potassium oxide (K_2O), and the Si/Al ratio of the source material, the type of alkaline activator that is used, the degree of dissolution of silicon etc. [67]. All these factors significantly influence the compressive strength of Geopolymer materials.

2.6.2. Alkaline activator and heat treatment properties

As mentioned above, increasing the concentration of the two components of the alkaline activator, (Na_2SiO_3 and NaOH), have a beneficial effect in terms of performance of the alkali-activated geopolymer material. Higher concentrations lead to the formation of much stronger chemical bonds, which results in a geopolymer matrix with superior properties [30,68]. Heat treatment temperature is one of the important factors also. It was found [69] that the temperature to which geopolymer materials are exposed is an accelerator for the geopolymerization reaction and it greatly affects the mechanical properties of the material, together with the time of heat treatment, and the type of alkaline activator. A higher treatment temperature, together with a greater exposure time to the given temperature, has been shown to lead to higher results in compressive strength [70]. However, too high temperature, can lead to the appearance of cracks, which have a negative effect on the properties of the material [71].

The alkaline activator plays a vital role in initiating the geopolymerisation process. In general, a strong alkaline environment is required to increase the surface hydrolysis of aluminosilicate particles present in the raw material while the concentration of chemical activator has a pronounced effect on the mechanical properties of geopolymers [72,73]. Gorhan and Kurklu [74] investigated the influence of NaOH solution on compression resistance of alkali-activated geopolymer mortars at the age of 7 days. Three different concentrations of NaOH (3, 6 and 9M) were used during laboratory investigations, while other parameters such as sand/ash and sodium silicate/NaOH remained constant. The optimal concentration of the NaOH solution that produced the highest compressive strength at the age of 7 days, 22.0 MPa, was the NaOH - 6 M. solution.

The mechanical properties of the geopolymer materials increase, in general, as the chemical properties of the alkaline activator increases [21-28,36-39,75,76]. The optimal amount of alkaline activator, as well as the ratio of its constituent elements, can vary depending on many parameters such as: the properties of the raw materials used, specific compositional ratios, treatment temperature, etc.

2.6.3. Mixing procedure

In general, the raw material, rich in aluminum and silicon is combined with the alkaline activator and mixed together to form the geopolymer binder [5,6,14-16,26-28,48]. However, other ways of mixing were studied by adopting sequences in which the NaOH solution was initially combined with the fly ash, and then gradually adding the required amount of Na_2SiO_3 solution [49].

2.7. Alkali-activated geopolymer materials applications

It is very well known that Portland cement and concrete are used in the construction industry for a very long time. Currently, the applications for which they are used include many types of structures, more and more advanced, which is due to many years of studying and optimizing this type of material. Alkali-activated geopolymer materials are materials with cementitious properties that could also be successfully used in the industry. Even though the characteristics and behavior of these materials is still studied, with only a few applications in some parts of the world, the results are opening new opportunities for the production of this type of material, and operation at a high level.

Originally developed by Davidovits, *precast materials* are the most common way to use alkali-activated geopolymers [77]. Factory production control - manufacturing processes and the control of the heat treatment – allows these materials to be developed to ensure that, from the point of view of the mechanical behavior, all of the conditions imposed by the specific regulations are met [78].

Blocks and bricks are very common building materials in the industry, and those produced using alkali-activated geopolymer materials have begun to become more and more common [79]. Their production can be considered an alternative to traditional ones, in terms of the energy consumed for their production and also the possibility of reducing greenhouse gas emissions. Traditional bricks require the consumption of a large amount of energy for their burning, which is why using alkali-activated geopolymer materials for this concept is more than appropriate.

Materials Research Forum LLC
https://doi.org/10.21741/9781644901533

The development of alkali-activated *geopolymer elements* was also studied by assimilation of the existing standards for concrete elements with Portland cement, although for their production heat treatment was still necessary [80]. Also, by studying parameters that affect the mechanical performance of reinforced elements, composite geopolymer materials can be developed. Studies have shown good behavior of fiber reinforced alkali-activated geopolymer concrete elements [81].

Also, research on alkali-activated geopolymer materials includes directions on the possibility of developing *geopolymer cement powder*. In the case of concrete, Portland cement powder is the viable option for the development of the industry, only by adding water and aggregates. Several researchers studied the production of geopolymer cement powder by crushing alkali-activated geopolymer paste and by adding water, to form a new geopolymer compound with increased properties – the so-called "water-based geopolymer" [82-85].

The technological transfer from the laboratory to the industry was successfully achieved in Australia, by inaugurating in 2013 the first building made entirely of this type of material. The University of Queensland has built a 4-storey building by successfully using prefabricated geopolymer materials. Also in Australia, Queensland, a new airport was inaugurated in 2014, where some portions of it were made of prefabricated elements of alkali-activated geopolymer concrete (pathways, aircraft parking, adjacent structures, etc.).

Chapter 3

Raw Materials Used in the Production of Alkali-Activated Geopolymer Materials

3.1. Why alkali-activation?

For the world economic system, cement and concrete are indispensable elements of the construction industry. Worldwide, cement production for 2008 was about 2.9 billion tons, making concrete the most produced material in the world [86]. Production of such a large volume of cement / concrete is directly associated with environmental problems - cement production being responsible for about 5-8% of the total carbon dioxide emissions [87]. Demand for concrete, hence for cement, is constantly growing, especially in highly developed countries [88], which means that alternative binders are urgently needed to meet the needs of millions of people, without compromising the CO_2 levels of the atmosphere.

Cement industry estimates that currently, with the best available technologies for the production of Portland cement, CO_2 emissions resulting from its production could decrease by a maximum of 17% (even through the use of alternative fuels, optimizing the amount of cement in concrete, recycling, etc.) [89].

Although cited often, but way too little, one of the major benefits of the alkali-activated geopolymer materials is the low values of carbon dioxide emissions associated with their production, because it avoids the process of annealing at a high temperature for the synthesis of the raw materials (ash, slag, etc.).

Despite the visible progress in reducing the environmental impact, the cement industry contributes to its pollution not only by carbon dioxide emissions, but also by exploiting natural mineral reserves, which in small countries in Europe or East Asia has a negative impact, because the demand for concrete is high and the available resources are limited [90].

At the same time, we must also take into account the fact that the production of activators based on sodium silicate and / or sodium hydroxide also produces carbon, but are not the "greenhouse effect" type of emissions (SO_x, NO_x, phosphates, etc.), This can also become a problem if rigorous methods for their production don't develop [91,92].

Although environmental issues are currently not sufficiently aware to create a sufficiently high demand for the production of geopolymer technology, this is expected to increase with the imposition of rules on carbon dioxide emissions and their effects.

Currently, two significant barriers are being discussed when new materials are to be introduced in the construction industry [90]:

- The *legislative framework*: the need for standards and normative, both on conditions and parameters of the production of the material, but also on their mechanical characteristics; their development and market introduction require a lot of time and is a gradual process;

- *Durability*: independent on the characteristics of load-bearing capacity and physical-mechanical strength, determined relatively quickly and easily on groups of related materials, an essential problem is the sustainability of such materials, taking into account the fact that they must satisfy certain requirements, they must withstand for a long time, and for new materials this is not quantified yet;

3.2. Porland cement and fly ash

Portland cement, an active hydraulic binder, is the most widely used building material in the world for manufacturing concrete. Portland cement is produced by extracting raw materials from quarries, crushing, grounding and then blending them in a mill (Fig. 3.1).

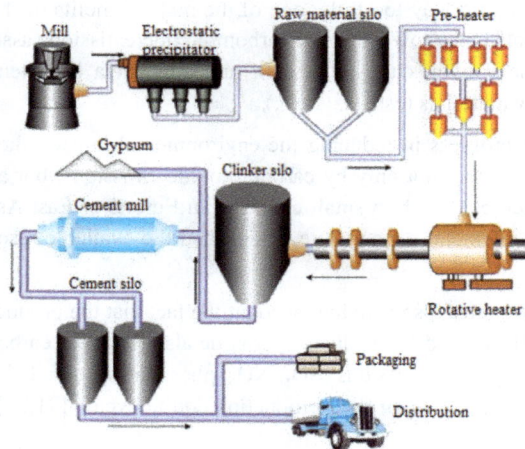

Figure 3.1. Production of Portland cement.

The main raw materials for producing cement are limestone (natural $CaCO_3$), clay or shale (SiO_2, Al_2O_3), Iron oxide (Fe_2O_3), silica sand (SiO_2) and gypsum ($CaSO_4$) [93]. Two distinct methods, wet and dry processes, are commonly used for the production of the clinker. However, more energy consumption and fuel-related costs are involved in wet processes rather than dry processes. Combustion is carried out at a temperature ranging from 1250 to 1500°C, and then the clinker is finely ground in the cement mill with some additives, until the average particle size is about 10 to 50 μm. An anhydrous gray powder, known as Portland cement, is stored in cement silos and is ready to be packed and shipped for use [94,95].

Portland cement acquires adhesive properties when mixed with water to become cement paste. The chemical reaction between cement and water is usually called cement hydration (Fig. 3.2). In general, hydration of cement can occur either by dissolution mechanism or by topochemical mechanisms. In the case of the dissolution mechanism, the cement dissolves in the solution. The resulting products are then precipitated in the form of hardened cement. The topochemical mechanism, sometimes called solid, obtains reactions directly on the surface of the cement as a hydrolysis rather than dissolution. However, both mechanisms could be involved in hydration, but due to a less solubility of cement, the reaction in the solid state is considered more favorable [95]. With multiple Portland cement compounds, hydration can be completed in different time periods or in different final properties, either one or both. Hydration products of the individual compound are similar to each other due to the conjugation of the main constituents, namely CaO, SiO_2 and Al_2O_3. In Portland cement, the approximate composition of hydrated calcium silicate (C-S-H: $C_3S_2H_3$) is formed with the assumption of the complete hydration process [96].

$$\text{Portland Cement} + \text{Water} \longrightarrow \text{Gel C-S-H} + Ca(OH)_2 + \text{Heat}$$

Figure 3.2. Cement paste production mechanism.

The additions in the cement or concrete are known as "a material other than water, aggregates or cement that are used either before or during mixing" [97], and the "material added during the mixing process of concrete in order to modify the properties of the mixture in the fresh state, and / or reinforced" [98].

Power plant ash (fly ash) can be defined as the fine dust, consisting mainly of spherical-shaped vitreous particles, from the combustion of pulverized coal [99,100]. It can be used in three directions (Fig. 3.3.).

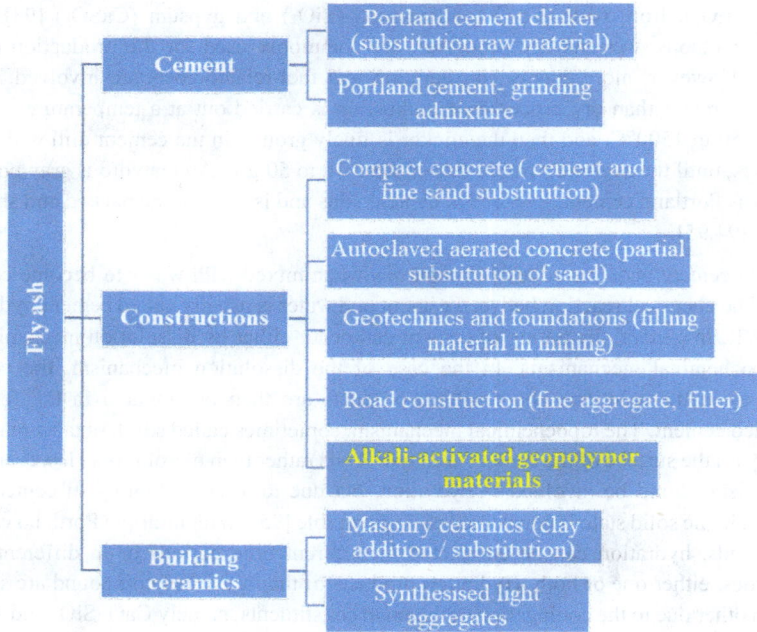

Figure 3.3.   Use of fly ash in the industry.

Fly ash resulting from the combustion of coal in power plants has two components: fly ash (consisting of particles which are discharged together with the flue gases) and ash collected from the hopper of the boiler (where it becomes unusable and is deposited in waste dumps in the vicinity of the power plant). Fly ash is generally captured by electrostatic filters before the flue gases reach the chimneys of the power plants. Pollution control equipment shall retain ash before the release of flue gases into the atmosphere.

The typical fly ash retention equipment is the electrostatic precipitators (Fig. 3.4). As a result of the burning of coal, the electrostatic boilers capture the fly ash. By pneumatic transport, it is transported in dry condition to specific silos in order to be distributed, depending on its destination (cement ash, concrete ash, etc.) (Fig. 3.5). Depending on the physical and chemical properties, fly ash is classified as cement-added ash (coarse fraction) and concrete-added fly ash, category A, N, being a Type II admixture for concrete production according to EN 450 [100].

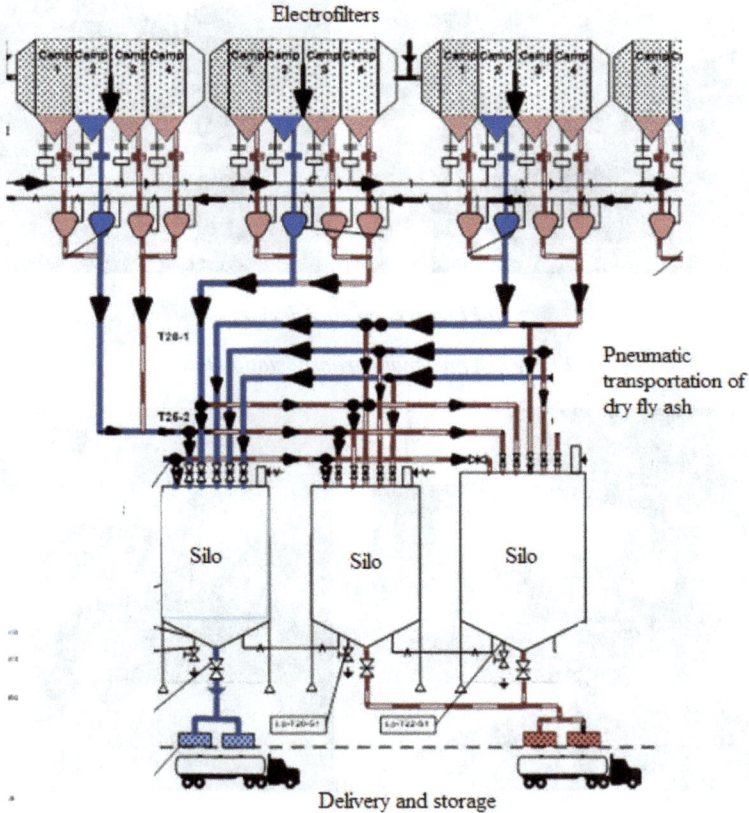

Figure 3.4.   Fly ash capturing process and its distribution.

a) Fly ash capturing device

b) Fly ash pneumatic transportation line

c) Fly ash collecting vessel

d) Mixing fly ash with water – 1:10

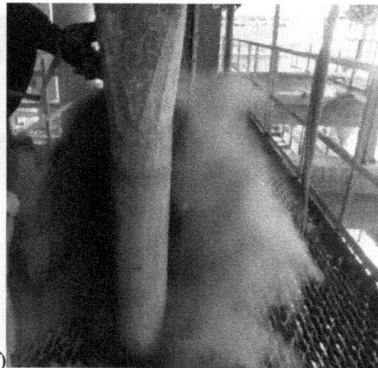

e) Fly ash electrostatic precipitator

f) Fly ash directly from the electrostatic precipitator

g) Fly ash distributing silo

Figure 3.5.   Power plant technological flow [101].

3.2.1. Fly ash for cement

Fly ash used in cement, is subject to the requirements laid down by EN 197-1, not falling within the scope of the EC marking [99]. Fly ash is obtained by electrostatic or mechanical deposition of powdery particles contained in the flue-gases from coal-fired boiler outbreaks (according to EN 197-1). As for the nature of the fly ash it can be siliceous (pozzolanic properties) or calcic (pozzolanic and hydraulic properties). Another important property of fly ash is the loss on ignition (L.O.I.), which is determined according to EN 196-2. At a calcination time of 1 hour, the L.O.I. should not be more than 5% of the mass. However, L.O.I. between 5 and 7% can also be accepted, the only condition being to comply with the durability conditions (freeze-thaw conditions and its compatibility with additives), the maximum value which should be stated on the package or on the final delivery document [102].

3.2.2. Fly ash for construction industry

Fly ash used in the construction industry, as type II addition in concrete, falls within the scope of EN 450-1 [100]. As regards the scope of the EC marking, it shall be subject to conformity assessment after System 1+. Because for the production of alkali-activated fly ash-based geopolymer concrete there is no standard in force, the type of fly ash used for the production of this material should imply research regarding the conditions of the fly ash according to existing standards and by analyzing some of fly ashes from Romania in order to determine which of these falls within the EN 450-1 standard, what are their possibilities for their certification and whether there are already certified.

3.2.3. Fly ash as type II addition for concrete

Type II addition means a finely divided, inorganic, pozzolanic or latent hydraulic material, which can be added to concrete to improve certain properties or to obtain special properties [100]. Fly ash is that fine powder consisting mainly of spherical vitreous particles, obtained by burning coal with or without co-burning materials, with outstanding pozzolanic properties, consisting mainly of SiO_2 and Al_2O_3. According to research in the field of geopolymer materials, existing fly ash, with high content of SiO_2 and Al_2O_3 is the necessary source material for the production of this type of concrete.

Worldwide, the amount of ash from coal burning is estimated to be about 700 million tons / year. Approximately 70% of this amount is represented by fly ash, optimal material for use as a pozzolanic material in cementitious systems [103]. Although there is a large amount of fly ash produced, about 80% of this amount is not used efficiently, it is used in low – value applications - earth fillers, base layer for road structures, but also mostly stored material [102]. Storage of fly ash in the form of waste is extremely harmful, as it contributes to both pollution of the land and pollution of the atmosphere and groundwater.

The total production of fly ash, at European level, is estimated to be around 250 million tons per year. Seven countries have the majority share for obtaining this quantity of ashes: Russia, Poland, Czech Republic, Romania, Germany, Spain and England [103]. The

European body which is involved in the study of coal combustion products (symbolized CCPs – Coal Combustion Products) is ECOBA (European Coal Combustion Products Association). CCPs are classified as it follows:

- FA – fly ash;

- BA – furnace bottom ash;

- BS – boiler slag;

- FBC – fluidized bed combustion ash;

- SDA – semi dry absorption product;

- FDG – flue gas de-sulphuration gypsum.

Recent estimations of ECOBA [104], indicate the share of different types of CCPs produced at European level. The European production of about 65 million tonnes per year of the 15 member states is represented in Fig. 3.6:

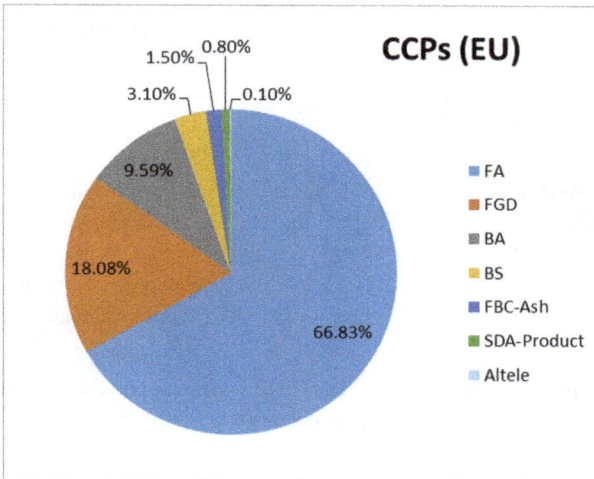

Figure 3.6.   Main CCP production (UE-2005).

It can be seen, according to Fig. 3.6, that the share of fly ash is the highest. As regards, the rate of use at European level of the amount of fly ash is approximately 48% of the total quantity obtained. The main area of use was in the construction industry, the largest amount being used as an addition in concrete (Fig. 3.7). Since January 1st, 2007, for the use of fly ash as an addition in concrete, the EC marking became mandatory.

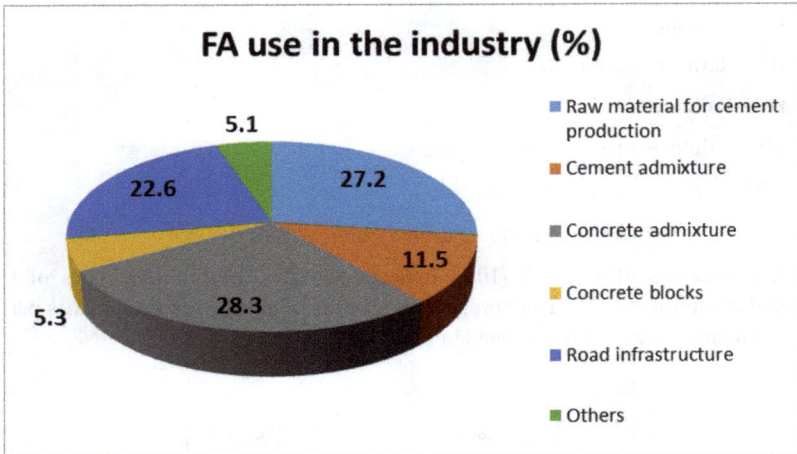

Figure 3.7.   Use of fly ash in the industry (2005).

In Romania, more than 80% of industrial waste is stored, which is the most used disposal method. By 2002, the number of industrial landfills in the country numbered about 680 warehouses, occupying more than 10000 ha. Of their total, 55 represent slag and ash dumps, 145 being tailing ponds / batals. Only 6 of these warehouses have all the facilities to meet the necessary conditions for compliance with the environmental protection norms in force [105,106].

Waste recovery in Romania, at the level of 2002 was extremely low. In total, only 5422 tons were recovered, representing approximately 0.06% of the total – an extremely small amount.

As regards the legislative framework in Romania, the Construction Products Directive DPC 89/106 is applied by Government Decision HG 622/2004. Although the total production of fly ash is more than 8 million tons per year, fly ash will not be used in the manufacture of concrete as an EC-labelled product subject to compliance under System 1+. The corresponding Romanian act of Directive 89/106 / EEC for establishing the conditions for placing on the market of construction products is the Government Decision

GD 622/2004 with the additions from GD 796/2005. This Government Decision rules the obligation to apply EC marking to construction products since February 27, 2005 [107].

In 2017 alone, a single thermal power plant in Romania produced almost 650,000 tons of ash, 50,000 tons of slag and 50,000 tons of gypsum. Although the recovery of the slag by recycling it back into the flow of the coal-fired power boilers (43.000 tons/year), and the delivery of fly ash to the construction industry (cement, ready-mixed concrete, mortars for soil stabilizing, roads, etc. - 162.000 tones/year), the disposal of the ash from the power plant is still an issue for the power plant – almost 565.000 tones/year in 2017 for ash and gypsum storage in waste dumps [101].

Although for the production alkali-activated fly ash-based geopolymer materials a certification of the used fly ash is not required, in order to be able to produce this type of concrete it is necessary to know the existing industry and all possibilities of harnessing the thermal power plant ash in Romania, as well as the choice of the other raw materials. In Romania, there are many thermal power plants whose ash, resulting as waste, could be exploited for the production of alkali-activated fly ash-based geopolymer materials.

3.2.4. General conditions for producing fly ash

Fly ash is the material obtained by electrostatic or mechanical precipitation of pulverizing particles resulting from the flue-gases of furnaces, supplied with pulverized coal. It can be processed by grading, selecting, sifting, drying, mixing, grinding with carbon reduction, as well as by combining these procedures [102]. It is well known that for the production of electricity, the use of coal leads to the generation of large amounts of fly ash. In order to be used successfully, fly ash can be subjected to processing, to optimize its fineness, to reduce the need for water or to improve other properties. They can be processed according to EN 450-1 standard, in which case it is referred to. When the ash is out of the scope of this standard, the possibility of using it as a type II addition in the concrete, or as a source material for the production of other types of concrete can be determined by the standards or other provisions such as National or European Technical Approvals.

For the production of alkali-activated fly ash-based geopolymer materials, fly ashes from Romania that may comply with EN 450-1 will be analyzed, although the production of this type of concrete does not require compliance with this standard, in order to produce a material that may comply with the current or future standards. Depending on the country-specific national experiences, the use of flying ash is strictly restricted after the dependence of loss on ignition parameter (L.O.I.) [102]. The European level trend is to bring the quality parameters of fly ash within the limits of the existing standards and to comply with the current requirements.

The population needs for energy continue to increase, which, at global level, leads to an increase in the production of fly ash, however, the issue of the quality of this type of waste is constantly increasing as the quantity of fly ash with low L.O.I. (loss on ignition) available on the market continues to decline. This is due to the new environmental requirements for NOx emissions and the fact that thermal power plants equipped with "low-NoX" burners have lower coal burning temperatures [108]. Although the new environmental protection rules should come to the environmental advantage of producing flying ash, they have a contrary effect on its quality.

The sale of fly ash is not the core business of a thermal power plant, which is a by-product, a waste product that results from the burning of coal for electricity generation, but it can also be a significant source of economic benefit for the production of income for the companies in the industry. The community is putting pressure on the compliance with the environmental standards and to reduce as much as possible the impact on the environment due to the waste that results from industrial processes, so it is desirable to minimize the production of waste, its recovery in terms of its use in other branches of industry, the destruction of their heating through the recovery of the used energy.

Therefore, the main objective of the majority of the electric power companies, with regard to the protection of the environment, is to establish an integrated management of by-products, by significantly reducing the amount of waste being stored, which is a pollutant, but an increase in the production of energy through non-polluting methods, and the recovery of waste by means of re-integration in industrial processes related to [108]. One of the important problems arising in the widespread use of fly is precisely the difficulty of controlling the quality of this by-product, which should meet certain standards in order to be used.

In Romania, following the process of burning fossil fuels, two types of wastes are obtained: fly ash (fine fractions, taken from electrostatic precipitators) and boiler slags (coarse fractions, resulting in the lower part of the boiler outbreak). The disposal of solid waste resulting from combustion is carried out hydraulically, the resulting material being transported to the storage dumps. Although this mechanism requires a number of technological steps, which can destroy the resulting ash in some power plants in the country, there are some power plants which are optimal for producing fly ash and delivering it to the industry (e.g. Govora Power Plant, Mintia Power Plant). However, it can be said that fly ash produced in coal-fired plants in Romania is not treated properly and on an industrial scale, it is not considered to be a product to which the conformity is evaluated according to the system of 1+, making it less available for the production of concrete according to the specific standards [109-111].

For the preliminary identification of locally available fly ash resources and the identification of relevant characteristics for the evaluation of their potential use for the development of alkali-activated geopolymer materials, a trip to CET. Govora was achieved in order to study and observe the process of production of industrial by-products related to the activity of the power-plant – namely the production of fly ash.

In this case, at Govora Thermal Power Plant, following the combustion process of fossil fuels, two types of waste are obtained: fly ash (fine fractions, taken from electro filters - which in turn are classified into several categories) and hotbed or boiler slags (coarse fractions, resulting in the lower part of the boiler hotbed). The discharge of solid waste resulting from combustion is carried out hydraulically, the resulting material being transported to the storage dumps. Although this mechanism involves a series of technological steps that can destroy the resulting ash, at CET Govora there are optimal facilities for dry ash capture, with its transport and delivery to cement plants.

3.2.5. *Classification of fly ash according to EN 450-1*

The conditions to be met both physically and chemically by fly ash in order to be used as Type II additions for concrete production are defined by Standard EN 450-1, revised and updated. The conditions regarding the composition, mixing, commissioning or storage of fly ash in concrete production however come outside the scope of this standard. For these, references to other standards in force for concrete will be made. Parameters to be met according to the old standard will be established, since the data from the Romanian thermoelectric power plants on fly ash characteristics were made according to the old standard, then they will be compared with the requirements of the 2012 standard.

Most power plants in Europe produce fly ash for concrete with a loss on ignition L.O.I. <5% [112]. Depending on the type of burned coal, the process to which it is subjected and certain operating conditions, flying ash may have different characteristics. The composition and behavior of fly ash change from one case to another, it is precisely because of this that their standardization is carried out difficult. In general, fly ash contains a variable amount of unburned carbon, which leads to a difference between each flying ash in terms of the L.O.I.

Since carbon can have a negative impact on concrete, a large part of concrete producers have imposed a L.O.I. permissible limit of up to 4%, although according to the current standards this value can be up to a maximum of 9% [108]. In relation to alkali-activated fly ash-based geopolymer materials, a larger percentage of the L.O.I. parameter does not influence any of the characteristics of the concrete because the chemical processes for their production are not similar at all to the production of ordinary Portland cement concrete

For the production of alkali-activated fly ash-based geopolymer materials, there is no harmonization of fly ash with the existing standards. In the literature, when specifying the materials used for the preparation of the mixes, fly ash characteristics are presented as if they were standardized, but not specifying whether or not they fall within certain limits.

The chemical (Table 3.1) and physical (Table 3.2) characteristics of the fly studied for their certification according to EN 450-1 are:

Table 3.1.   Limits imposed by EN 450-1 + A1 for the chemical characteristics of fly ash.

Element	Limit	Limit description
Loss on ignition (L.O.I.)	**Category A**: <5% by mass **Category B**: 2.0%↔7.0% by mass **Category C**: 4.0%↔9.0% by mass	Value range
Chloride	0.10% by mass	Maximum value
Sulphuric anhydride (SO$_3$)	3.0% by mass	Maximum value
Free CaO	2.5% by mass	Maximum value
Reactive calcium oxide	10.0% by mass	Maximum value
Reactive silicon dioxide	25% by mass	Maximum value
Silicon dioxide (SiO$_2$), aluminum oxide (Al$_2$O$_3$) and iron oxide (Fe$_2$O$_3$)	Sum – 70% by mass	Maximum value
Total alkali content	5.0% by mass	Maximum value
Magnesium oxide (MgO)	4.0% by mass	Maximum value
Soluble phosphate (P$_2$O$_5$)	100 mg/kg	Maximum value

Table 3.2.   *Limits imposed by EN 450-1 for the physical characteristics of fly ash.*

Characteristic	Limit	Limit description
Fineness	*Category N*: 40% retained fly ash (by mass); ±10 % variation limits from declared value; *Category S*: 12% retained fly ash (by mass);	Maximum value on the 0,045mm sieve
Activity index	75% at 28 days 85% at 90 days	Minimum values
Expansion	10 mm	Maximum value
Density	± 200 kg/m^3	Maximum deviation

Table 3.3.   *Standards adopted to analyse the physical and chemical characteristics of fly ash.*

Characteristic		Analysis method
Loss on ignition (%)	Chemical properties	EN 196-2
Cl (%)		EN 196-2
SO$_3$ (%)		EN 196-2
Reactive CaO (%)		EN 196-2
Reactive SiO$_2$		EN 196-2
SiO$_2$ + Al$_2$O$_3$ + Fe$_2$O$_3$ (%)		EN 196-2
Total alkali content (%)		EN 196-2
MgO (%)		EN 196-2
Free CaO (%)		EN 451-1
Soluble phosphate P$_2$O$_5$ (%)		EN 450-1
Fineness (R$_{0.045}$) (%)	Physical properties	EN 451-2
Activity index (%): - 28 days - 90 days		EN 196-1
Stability (mm)		EN 196-3+A1
Initial setting time (minutes)		EN 196-3+A1
Density (kg/m^3)		EN 196-6
Water requirement (%) – only for category S fly ash		EN 450-1+A1

The analysis of standard EN 450-1 revealed that Annex ZA contains 18 important characteristics that fly ash must meet in order to be used in concrete. For the certification of fly ash for its use in concrete, both their physical and chemical characteristics must be known. For this purpose it is necessary to establish their conformity or non-compliance with the acceding criteria and to establish a degree of compliance with these technical conditions imposed by the standards presented above in Table 3.3.

Chapter 4

Development of Alkali-Activated Geopolymer Binder

In detail data about all the processes adopted to produce alkali- activated fly ash-based geopolymer binder - called alkali-activated geopolymer paste (AAGP) - will be presented in this chapter. The chapter aims to establish preliminary alkali-activated fly ash-based geopolymer paste mixtures, to optimize them and to study the influence of different salient parameters that affect the material.

To simplify the process of developing these types of materials, compressive strength was chosen as reference parameter, because of its fundamental importance in the design of cementitious materials in the construction industry.

Although the alkali-activated geopolymer materials can be produced using different types of source materials, in the current research low-calcium fly ash was used. The mix design procedures have been based on the study of the literature and on the principles and standards related to the production/testing of ordinary Portland cement.

4.1. General principles for the production of alkali-activated geopolymer materials

The alkaline activity of each component of the geopolymer paste affects both its properties in the fresh state and its properties in the hardened state. In order to produce alkali-activated geopolymer materials, it is advisable to take into account the variations that occur from one casting to another, and to try to minimize the possible differences in the results. It is optimal to perform a minimum of two or three tests for each mixture, for parametric confirmation of the expected performance, regime and requirements. Alternatively, in order to avoid the failure characteristics that are determined for the final product and to ensure its "self-sufficiency", it is recommended to use a moderate oversize of some of the critical properties of the mix-design [113].

Most of the existing studies on geopolymer materials are based on the production of geopolymer paste and geopolymer mortars, their properties being measured on relatively small size samples. Moreover, the full details about the mix-design are not presented. Geopolymerization is a chemical process, and in order to be able to establish mix-design directions, it is mainly necessary to take into account the final chemical composition of the mixtures, based on the properties of the raw materials.

Materials Research Forum LLC
https://doi.org/10.21741/9781644901533

As a new material, dependent on a number of factors that influence it, every researcher that is studying this type of material adjusts the mix-design of the alkali-activated fly ash-based paste based on the local raw materials. In order to be able to develop the technology for producing alkali-activated geopolymer paste, a rigorous "trial-and-error" process was considered to be the best approach in determining the factors that influence this type of material. One of the main objectives of this phase is the mix-design and production of preliminary geopolymer paste and concrete mixes, as well as the design of an adapted laboratory technology, by:

-     establishing sequences for producing NaOH solution;

-     making the alkaline solution by combining NaOH solution and Na_2SiO_3 solution;

-     establishing the mixing sequences, as well as the order of mixing of the materials and the optimal mixing time;

-     identifying the proper heat curing treatment;

-     identifying the proper conditioning after demoulding.

The aim is to ensure a proper geopolymerization process, as well as good workability and homogeneity of the developed mixtures.

As far as possible, the current practice used for the production and testing of ordinary Portland cement concrete has been pursued. Using the established methodology was considered timely and functional, by referring to classical materials and also to facilitate the promotion of this "new" material in the industry. To simplify the development process of these types of materials, compressive strength was selected as a reference parameter. This is not unusual, since compressive strength has major importance in the structural design of buildings.

The main objectives for the development of AAGP were based on the following criteria:

-     Preliminary identification of locally available fly ash resources and identification of relevant characteristics to assess their potential use in the development of AAGM;

-     Establishing the specific processes for the production of AAGP;

-     Evaluation of the effect of the mixing sequence between the alkaline activator and the fly ash in the mixtures;

-     Assessment of the behavior of fresh alkali-activated geopolymer paste mixtures;

-     Developing the mixing process and establishing a proper heat curing regime;

-     Optimising the basic mixture ratios of the alkali-activated fly ash-based geopolymer paste.

4.2. Alkali-activated fly ash-based geopolymer paste – raw materials characteristics

4.2.1. Fly ash (FA)

Three viable Romanian fly ash sources were identified for the production of alkali-activated fly ash-based geopolymer paste:

- Govora Power Plant, Vâlcea County – fly ash G1 and G2 (Fig. 4.1 and Fig. 4.2);

- Mintia Power Plant, Hunedoara County – fly ash M1 and M2 (Fig. 4.3 and Fig. 4.4);

- Rovinari Power Plant, Gorj County – fly ash R (Fig. 4.5).

Those were obtained from the electrofilters from each power plant and their chemical composition was obtained by XRF (X-Ray Fluorescence) analysis and is presented in Table 4.1.

Figure 4.1.   Fly ash G1.

Figure 4.2.   Fly ash G2.

Figure 4.3.   Fly ash M1.

Figure 4.4.   Fly ash M2.

Figure 4.5.   Fly ash R.

Table 4.1.   Chemical composition of fly ash samples (% by mass).

Oxides	Fly ash G1	Fly ash G2	Fly ash M1	Fly ash M2	Fly ash R
SiO_2	51.76	54.32	53.75	53.61	46.94
Al_2O_3	21.86	22.04	26.02	26.16	23.83
Fe_2O_3	9.40	9.02	7.91	7.58	10.08
CaO	6.56	5.85	2.54	2.42	10.72
MgO	2.43	2.48	1.54	1.49	2.63
SO_3	0.38	0.19	0.35	0.26	0.45
Na_2O	0.37	0.54	0.59	0.59	0.62
K_2O	2.16	2.23	2.57	2.60	1.65
P_2O_5	0.14	0.16	0.12	0.12	0.25
TiO_2	0.84	0.86	1.02	1.04	0.92
Cr_2O_3	0.04	0.03	0.05	0.05	0.02
Mn_2O_3	0.04	0.06	0.09	0.08	0.06
ZnO	0.03	0.02	0.04	0.03	0.02
SrO	0.03	0.02	0.02	0.02	0.03
L.O.I.*	3.69	3.05	3.14	3.57	2.11

It can be seen from Table 4.1 that all fly ash samples have low L.O.I. values, which means that they have low carbon content. Also, all of the samples have a Si/Al molar ratio of approximately 2, which makes them suitable for the production of alkali-activated geopolymer paste, due to the Si-O-Al-Si bonds that they could provide and develop further in the mixture.

All five samples fly ash qualify as being Class F fly ash [114]. The data is presented in Table 4.2:

Also, for all five samples, fineness parameter R0.045 (residue retained on the 0.045 mm sieve) was determined in order to establish their physical characteristics (Table 4.3 and Fig. 4.6).

Table 4.2. Class F compliance of fly ash samples.

Fly ash sample	Chemical characteristics		
	$SiO_2 + Al_2O_3 + Fe_2O_3$ [>70%]	SO_3 [<5%]	L.O.I. [<6%]
Fly ash G1	83.02	0.38	3.69
Fly ash G2	85.38	0.19	3.05
Fly ash M1	87.68	0.35	3.14
Fly ash M2	87.35	0.26	3.57
Fly ash R	80.85	0.45	2.11

Table 4.3. Fineness $R_{0.045}$ of the fly ash samples (% by mass).

Fly ash sample	Fineness $R_{0.045}$
Fly ash G1	59.40%
Fly ash G2	51.60%
Fly ash M1	39.20%
Fly ash M2	37.60%
Fly ash R	31.40%

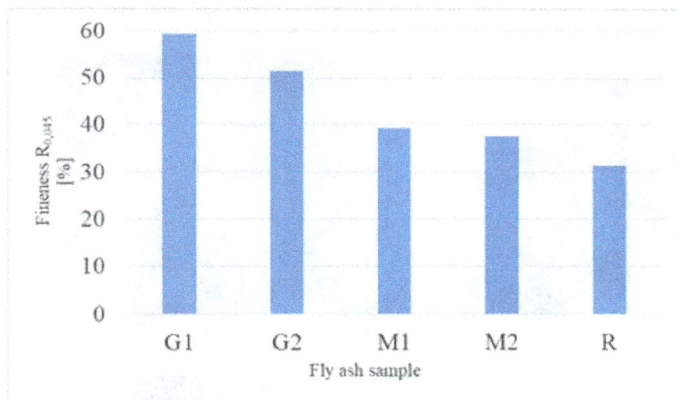

Figure 4.6. Fineness $R_{0.045}$ of the fly ash samples.

4.2.2. Alkaline activator (AA)

For the production of the alkali-activated geopolymer paste mixtures mixtures, a combination between sodium hydroxide solution of different concentrations (NaOH solution) and sodium silicate solution (Na_2SiO_3 solution) was chosen as alkaline activator.

Sodium hydroxide solution (NaOH)

In general, sodium hydroxide (NaOH) is purchased commercially, in solid form (pallets) or as flakes or pearls (Fig. 4.7 and Fig. 4.8). The cost of its purchase varies depending on the purity of the substance.

The sodium hydroxide solution (SH) was produced by dissolving the NaOH flakes or pearls in water until they were completely dissolved. The preparation of NaOH solution is an exothermic reaction (the reaction occurs with heat release) and, therefore, certain laboratory rules and regulations are applied, which fall under the scope of Regulation EC 1272/2008. Sodium hydroxide is a corrosive material to skin (category 1A), resulting in severe burns to the skin and damage the eyes in case of contact [115]. It is advised never to add water to the corrosive substance when producing the solution. The addition of sodium hydroxide flakes into the water was achieved in small quantities, gradually. To avoid excessive heat generation, cold water was used at a temperature of not more than 20°C. Afterwards, NaOH solution was stored in sealed packaging, under laboratory conditions, to avoid contact with air, moisture and carbon dioxide

Figure 4.7.   NaOH flakes.

Figure 4.8.   NaOH pearls.

The main characteristic of sodium hydroxide solution is its *molarity*. It can be different from one solution to another. This feature has an important influence on the final strength of alkali-activated fly ash-based geopolymer materials, as highlighted in the previous chapter.

The molar concentration of the sodium hydroxide solution is characterized by the number of NaOH moles present in the aqueous solution. Thus, in the research conducted, three molar NaOH solution concentrations were used, established as following:

8M = 8x40g = 320g NaOH flakes or pearls / 1 liter of solution

        1kg NaOH 8M solution = 262g NaOH flakes or pearls

10M = 10x40g = 400g NaOH flakes or pearls / 1 liter of solution

        1kg NaOH 10M solution = 314g NaOH flakes or pearls

12M = 12x40g = 480g NaOH flakes or pearls / 1 liter of solution

        1kg NaOH 12M solution = 361g NaOH flakes or pearls

The amount of flakes or pearls of sodium hydroxide in the solution is expressed for 1 kilogram of solution, since in the design of the mixtures, the alkaline liquid is calculated based on mass. The mass of NaOH flakes represents only a fraction of the mass of the NaOH solution, water being the major component (Table 4.4). Establishing these parameters is very important, because they will generate the total amount of water released in the mixtures.

Table 4.4.   NaOH solution part distribution.

NaOH Concentration	Solid part [%]	Liquid part [%]
8M	26.23	73.77
10M	31.37	68.63
12M	36.09	63.91

Sodium silicate solution (Na$_2$SiO$_3$)

Sodium silicate solution (Na$_2$SiO$_3$) is also known as water-glass and is in liquid state. Its chemical properties are limited to the Na$_2$O to SiO$_2$ ratio. These chemical properties are provided by companies producing sodium silicate solution. In this research, three types of sodium silicate solution (SS) were used; their chemical characteristics are illustrated in Table 4.5.

Table 4.5.   Chemical characteristics of the used Na_2SiO_3 solutions (% by mass).

Type of Na_2SiO_3 solution	Na_2O [%]	SiO_2 [%]	H_2O [%]
SS1	9	31	60
SS2	14	30	56

Preparation of the alkaline activator

For all alkali-activated geopolymer paste mixtures, the procedure for mixing the sodium silicate solution with the sodium hydroxide solution followed the next steps:

-    Sodium hydroxide solution (NaOH solution) with specific molar concentration was prepared (8M, 10M or 12M);

-    Based on the Na_2SiO_3/NaOH pre-established ratio, the two solutions were mixed together after 24 hours, under laboratory conditions (T (20±1)°C and 60% RH);

-    After 24 more hours, based on the proposed mix-design, the mix between the alkaline activator and the fly ash was realized.

4.3. Alkali-activated fly ash-based geopolymer paste technology

As in the case of ordinary Portland cement mortars, a paddle mixer, with 4.75 l capacity, a rotational movement of 140/280 rot/min and 61.5/123 rot/min revolution, was used to produce alkali-activated fly ash-based geopolymer paste [116].

Mixing technology, specific to low volume experimental casting (maximum 0.8-1.0 l) included the following sequences (Fig. 4.10):

-    Conditioning of the raw material (fly ash) at (20±2)°C and preparation of the alkaline activator by combining the Na_2SiO_3 and NaOH solutions and verifying its temperature;

-    The fly ash was placed in the mixing container;

-    The mixing was started at low speed and the alkaline liquid was gradually added over the fly ash during 90s;

-    After the complete addition of the alkaline liquid over the fly ash, the mixing was continued at a low speed for 3 minutes, while the workability of the mixture was observed;

-    The mixtures were then placed in 40 mm x 40 mm x 160 mm molds and heat cured at 70°C for 24 hours. A glass film was placed on top of every mold in order to prevent excessive water release from the mixtures;

-    After demolding, the specimens were stored in the climatic chamber at the temperature T (20±1)°C and relative humidity RH (60±5)% until the age of 7 days, when their mechanical performances were tested (Fig. 4.9).

There are a lot of factors that influence alkali-activated fly ash-based geopolymer paste in order to produce a homogeneous mixture:

- alkaline activator to fly ash ratio;

- Na_2SiO_3 to NaOH solution ratio;

- completion of the chemical reactions between alkaline liquid compounds;

- the speed of adding alkaline liquid to the mixture.

Figure 4.9.   Demoulded geopolymer paste samples.

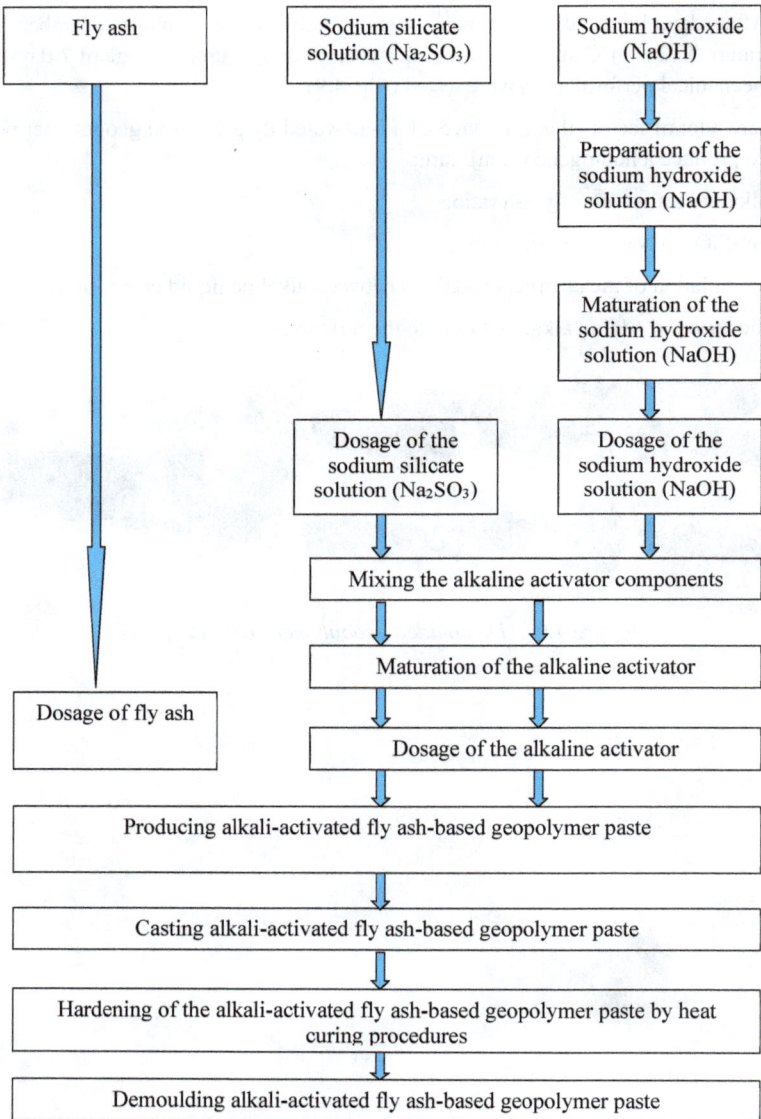

Figure 4.10. Preparation of the alkali-activated fly ash-based geopolymer binder.

4.4. Alkali-activated fly ash-based geopolymer paste characteristics

4.4.1. Alkali-activated geopolymer paste mixtures (AAGP)

Using the previously presented raw materials, a number of 47 geopolymer binder mixtures were produced, characterized compositionally and technologically in Table 4.6, respectively, in Table 4.7 from their oxide characteristics.

Table 4.6. Compositional and technological characterization of alkali activated fly ash-based geopolymer binder.

Mixture (Fly ash type)	Fly ash	Ratio to fly ash					Heat treatment
		AA/FA	SS / SH	SS (Na$_2$SiO$_3$)	SH (concentration)	Extra water	
GPP1 (G1)		0.66	0.50	0.20 (SS1)	0.30 (8M)	0,2	60°C / 24h
GPP2 (M1)			0.50	0.20 (SS1)	0.30 (8M)	-	
GPP3 (M1)			1.00	0.30 (SS1)	0.30 (8M)	-	
GPP4 (M1)			1.50	0.30 (SS1)	0.20 (8M)	-	
GPP5 (M1)			1.00	0.30 (SS2)	0.30 (8M)	-	
GPP6 (M1)			1.50	0.30 (SS2)	0.20 (8M)	-	
GPP7 (M1)			2.00	0.30 (SS2)	0.20 (8M)	-	
GPP8 (M1)			2.50	0.40 (SS2)	0.10 (8M)	-	
GPP9 (M1)			1.00	0.30 (SS2)	0.30 (8M)	0.10	70°C / 24h
GPP10 (M2)			1.00	0.30 (SS2)	0.30 (8M)	0.10	
GPP11 (M1)			1.00	0.30 (SS2)	0.30 (8M)	0.10	
GPP12 (M2)			1.00	0.30 (SS2)	0.30 (8M)	0.10	
GPP13 (M2)			1.00	0.30 (SS2)	0.30 (8M)	0.03	
GPP14 (M2)	1.00	0.50	0.50	0.17 (SS2)	0.33 (8M)	-	
GPP15 (M2)			1.00	0.25 (SS2)	0.25 (8M)	-	
GPP16 (M2)			1.50	0.30 (SS2)	0.20 (8M)	-	
GPP17 (M2)			2.00	0.33 (SS2)	0.17 (8M)	-	
GPP18 (M2)			2.50	0.36 (SS2)	0.14 (8M)	-	
GPP19 (M2)			0.50	0.17 (SS2)	0.33 (10M)	-	
GPP20 (M2)			1.00	0.25 (SS2)	0.25 (10M)	-	
GPP21 (M2)			1.50	0.30 (SS2)	0.20 (10M)	-	
GPP22 (M2)			2.00	0.33 (SS2)	0.17 (10M)	-	
GPP23 (M2)			2.50	0.36 (SS2)	0.14 (10M)	-	
GPP24 (M2)			0.50	0.17 (SS2)	0.33 (12M)	-	
GPP25 (M2)			1.00	0.25 (SS2)	0.25 (12M)	-	
GPP26 (M2)			1.50	0.30 (SS2)	0.20 (12M)	-	
GPP27 (M2)			2.00	0.33 (SS2)	0.17 (12M)	-	
GPP28 (M2)			2.50	0.36 (SS2)	0.14 (12M)	-	

Materials Research Forum LLC
https://doi.org/10.21741/9781644901533

Mixture (Fly ash type)	Fly ash	Ratio to fly ash					Heat treatment
		AA/FA	SS / SH	SS (Na2SiO3)	SH (concentration)	Extra water	
GPP29 (G2)			0.50	0.33 (SS2)	0.67 (8M)	-	
GPP30 (G2)			1.00	0.50 (SS2)	0.50 (8M)	-	
GPP31 (G2)			1.50	0.60 (SS2)	0.40 (8M)	-	
GPP32 (G2)			2.00	0.67 (SS2)	0.33 (8M)	-	
GPP33 (G2)			2.50	0.71 (SS2)	0.29 (8M)	-	
GPP34 (G2)			0.50	0.33 (SS2)	0.67 (10M)	-	
GPP35 (G2)			1.00	0.50 (SS2)	0.50 (10M)	-	
GPP36 (G2)		1.00	1.50	0.60 (SS2)	0.40 (10M)	-	
GPP37 (G2)			2.00	0.67 (SS2)	0.33 (10M)	-	
GPP38 (G2)	1.00		2.50	0.71 (SS2)	0.29 (10M)	-	70°C / 24h
GPP39 (G2)			0.50	0.33 (SS2)	0.67 (12M)	-	
GPP40 (G2)			1.00	0.50 (SS2)	0.50 (12M)	-	
GPP41 (G2)			1.50	0.60 (SS2)	0.40 (12M)	-	
GPP42 (G2)			2.00	0.67 (SS2)	0.33 (12M)	-	
GPP43 (G2)			2.50	0.71 (SS2)	0.29 (12M)	-	
GPP44 (R)			1.00	0.50 (SS2)	0.50 (8M)	-	
GPP45 (R)		0.90	1.00	0.45 (SS2)	0.45 (8M)	-	
GPP46 (R)		0.85	1.00	0.43 (SS2)	0.43 (8M)	-	
GPP47* (R)		*0.75*	*1.00*	*0.33 (SS2)*	*0.33 (8M)*	*-*	

*Note: Mixture GPP47 was produced using fly ash with $R_{0.045}=0\%$.

Table 4.7. *Oxidic characterization of alkali activated fly ash-based geopolymer binder*

Mixture (Fly ash)	Moles / mixture				Molar ratio		
	Na_2O	SiO_2	Al_2O_3	H_2O	Na_2O/SiO_2	SiO_2/Al_2O_3	H_2O/Na_2O
GPP1 (G1)	2.21	11.71	2.57	44.94	0.19	4.55	20.35
GPP2 (M1)	2.29	15.69	4.08	32.50	0.15	3.84	14.20
GPP3 (M1)	1.02	8.19	2.04	15.52	0.12	4.01	15.18
GPP4 (M1)	0.95	8.83	2.04	15.08	0.11	4.11	15.89
GPP5 (M1)	1.48	10.19	2.55	18.84	0.15	3.99	12.734
GPP6 (M1)	1.43	10.44	2.55	18.19	0.14	4.09	12.729
GPP7 (M1)	1.40	10.61	2.55	17.75	0.13	4.16	12.725
GPP8 (M1)	1.37	10.73	2.55	17.43	0.12	4.20	12.722
GPP9 (M1)	0.89	6.12	1.53	13.53	0.15	3.99	15.237
GPP10 (M2)	0.89	6.10	1.54	13.53	0.15	3.96	15.237

Mixture (Fly ash)	Moles / mixture				Molar ratio		
	Na_2O	SiO_2	Al_2O_3	H_2O	Na_2O/SiO_2	SiO_2/Al_2O_3	H_2O/Na_2O
GPP11 (M1)	0.89	6.12	1.53	13.53	0.15	3.99	15.237
GPP12 (M2)	0.89	6.10	1.54	13.53	0.15	3.96	15.237
GPP13 (M2)	1.55	8.47	2.05	21.21	0.18	4.13	13.66
GPP14 (M2)	1.56	9.75	2.57	19.9	0,16	3.8	12.742
GPP15 (M2)	1.48	10.17	2.57	18.8	0,15	3.96	12.734
GPP16 (M2)	1.43	10.42	2.57	18.2	0,14	4.06	12.729
GPP17 (M2)	1.39	10.59	2.57	17.8	0,13	4.13	12.725
GPP18 (M2)	1.37	10.7	2.57	17.4	0,13	4.17	12.722
GPP19 (M2)	1.78	9.75	2.57	19.2	0,18	3.8	10.793
GPP20 (M2)	1.64	10.17	2.57	18.3	0,16	3.96	11.15
GPP21 (M2)	1.56	10.42	2.57	17.7	0,15	4.06	11.394
GPP22 (M2)	1.5	10.59	2.57	17.4	0,14	4.13	11.572
GPP23 (M2)	1.46	10.7	2.57	17.1	0,14	4.17	11.707
GPP24 (M2)	1.98	9.75	2.57	18.5	0,2	3.8	9.376
GPP25 (M2)	1.79	10.17	2.57	17.8	0.18	3.96	9.945
GPP26 (M2)	1.68	10.42	2.57	17.3	0.16	4.06	10.349
GPP27 (M2)	1.6	10.59	2.57	17	0.15	4.13	10.648
GPP28 (M2)	1.55	10.7	2.57	16.8	0.14	4.17	10.881
GPP29 (G2)	3.03	10.7	2.16	39.9	0.28	4.95	13.177
GPP30 (G2)	2.86	11.54	2.16	37.7	0.25	5.34	13.194
GPP31 (G2)	2.75	12.04	2.16	36.4	0.23	5.57	13.206
GPP32 (G2)	2.69	12.37	2.16	35.5	0.22	5.72	13.214
GPP33 (G2)	2.64	12.61	2.16	34.9	0.21	5.83	13.22
GPP34 (G2)	3.45	10.7	2.16	38.4	0.32	4.95	11.116
GPP35 (G2)	3.18	11.54	2.16	36.6	0.28	5.34	11.512
GPP36 (G2)	3.01	12.04	2.16	35.5	0.25	5.57	11.785
GPP37 (G2)	2.9	12.37	2.16	34.8	0.23	5.72	11.983
GPP38 (G2)	2.82	12.61	2.16	34.2	0.22	5.83	12.136
GPP39 (G2)	3.85	10.7	2.16	37.1	0.36	4.95	9.628
GPP40 (G2)	3.47	11.54	2.16	35.6	0.3	5.34	10.241
GPP41 (G2)	3.25	12.04	2.16	34.7	0.27	5.57	10.678
GPP42 (G2)	3.1	12.37	2.16	34.1	0.25	5.72	11.003
GPP43 (G2)	2.99	12.61	2.16	33.7	0.24	5.83	11.257
GPP44 (R)	2.87	10.31	2.34	37.7	0.28	4.41	13.135
GPP45 (R)	2.59	10.06	2.34	33.9	0.26	4.3	13.084
GPP46 (R)	2.45	9.93	2.34	32	0.25	4.25	13.055
GPP47 (R)	1.9	9.43	2.34	24.5	0.2	4.04	12.893

4.4.2. Physical-mechanical properties of the alkali-activated geopolymer paste mixtures

An important aspect in the development of alkali-activated geopolymer paste is *workability*, which can, or cannot be obtained in mixtures when, during the geopolymerization process, water is released. It is important to take into account any design of the mixture and, therefore, to record at any time the behavior of fresh mixtures, so that subsequently they can be optimized.

The general observation was that when the $Na_2SiO_3/NaOH$ solution ratio increased, the workability of the mixtures also increased. Instead, the functionality of the mixtures with NaOH solution 10M and 12M was considerably lower than those using 8M NaOH solution. The high content of NaOH solids in the solution directly affects the workability of the mixes. This does not mean that NaOH is the only parameter that affects the functionality of the fresh fly ash-based geopolymer paste, however it is of great importance.

An important observation during the preparation of the samples was their behavior during and after mixing. For the apparently dry mixtures, the ones with high $Na_2SiO_3/NaOH$ ratios, at the beginning of the mixing the material was not homogenous. The term segregation is used in the case of alkali-activated fly ash-based geopolymer materials to describe the formation of material agglomerations in the mixture (2-3 cm "lumps "). This happens sometimes due to the sodium silicate solution in the mixture, which acts as an adhesive and "glues" together aggregates, liquids and fly ash. Therefore, for many of the mixtures with increased $Na_2SiO_3/NaOH$ ratios, the mixture seemed to be an almost completely dry mixture.

However, a significant aspect of these dry mixtures was that by vibrations, by using a shock mass to compact the samples, the mixtures formed a continuous and cohesive mixture. From an alkali-activated geopolymer paste sample that, to the naked eye, looked as if it was not compact, by vibrating, the mixture became reliable enough to be used practically. Moreover, it was noted that in general a drier mixture will generate higher compressive strength.

On the other hand, with low $Na_2SiO_3/NaOH$ ratios, geopolymer paste mixtures were very cohesive, with good workability. This is quite different from the usual conventional concrete, which does not show this kind of cohesiveness. High cohesiveness is again attributed to the action of the sodium silicate solution in the mixture that creates the "adhesion". It is also attributed to fly ash, where with little water content it is non-adhesive, but becomes cohesive by applying vibrations.

Hardened state apparent density of the alkali-activated fly ash-based geopolymer paste mixtures was determined according to EN 12390-7: *Testing hardened concrete. Density of hardened concrete* [117]. The volume of the test specimens was calculated by measuring the predetermined 40 x 40 x 160 mm prisms. The mass of the specimen was measured, in dry state (after conditioning 6 days at 20°C), under laboratory conditions, and the density was then calculated.

The evaluation of the apparent density is presented for the following mixtures, prepared with fly ash G2 and an alkaline activator prepared using SS2 sodium silicate solution and 8M, 10M and 12M, respectively, sodium hydroxide solution.

Results are presented in Table 4.8 for all analyzed samples and visually in Fig. 4.11 in order to present the influence of the NaOH solution concentration and the Na_2SiO_3 / NaOH solution ration on this parameter.

Note: In order to simplify the presentation of the obtained results, al the alkali-activated fly ash-based geopolymer paste samples have been grouped in several categories, based on the primary raw materials that were used in their production (Eq. 4.1).

AAGP FA αx SH yM SSz   (4.1)

Where: *AAGP* – alkali-activated geopolymer paste; *FA* – fly ash-based; *αx* – type of fly ash; *SH yM* – sodium hydroxide solution with y molarity; *SSz* – type of sodium silicate solution.

Table 4.8. *Apparent density in hardened state (7 days)*

Mixture	Apparent density [kg/m^3]
AAGP FA.G2 SH 8M SS2	
GPP29	1230
GPP30	1280
GPP31	1300
GPP32	1400
GPP33	1340
AAGP FA.G2 SH 10M SS2	
GPP34	1300
GPP35	1280
GPP36	1320
GPP37	1310
GPP38	1320
AAGP FA.G2 SH 12M SS2	
GPP39	1330
GPP40	1300
GPP41	1340
GPP42	1340
GPP43	1310

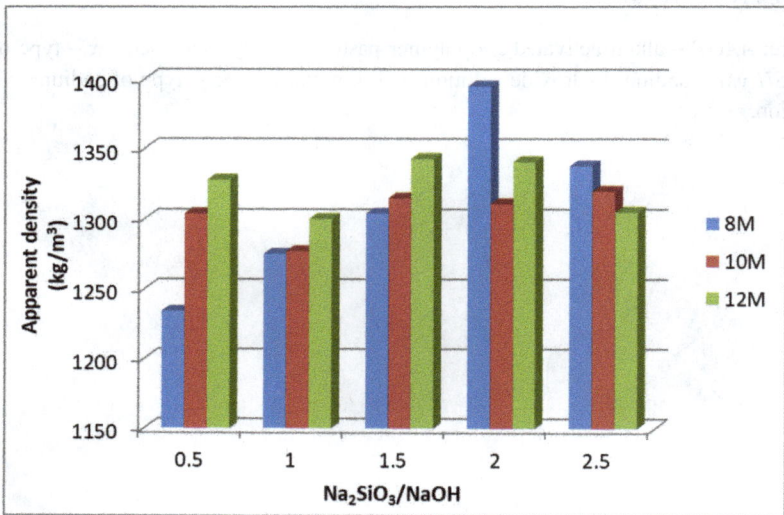

Figure 4.11. *Apparent density for AAGP FA.G2 SS2.*

Materials Research Forum LLC
https://doi.org/10.21741/9781644901533

Flexural strength of the alkali-activated geopolymer paste samples was determined by adopting the three point bending (3PB) test, according to EN 196-1 [116] and the procedure is presented in Fig. 4.12.

The samples were tested at the age of 7 days and the results are presented in Fig. 4.13.

For an easier comparison, Fig. 4.14 and Fig. 4.15 were chosen only for mixtures which were produced using the same type of fly ash.

a) . Placing the AAGP specimen in the equipment and failure under load

b) Visual evaluation of the specimens after flexural failure

Figure 4.12. 3PB flexural strength test.

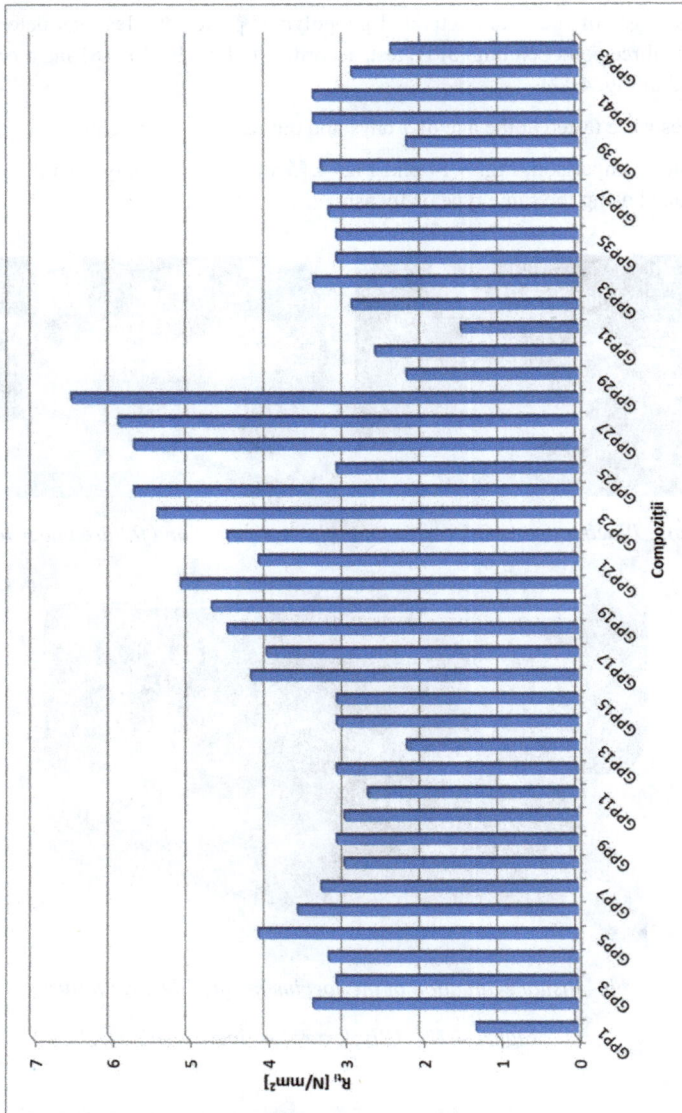

Figure 4.13. AAGP – 3PB results.

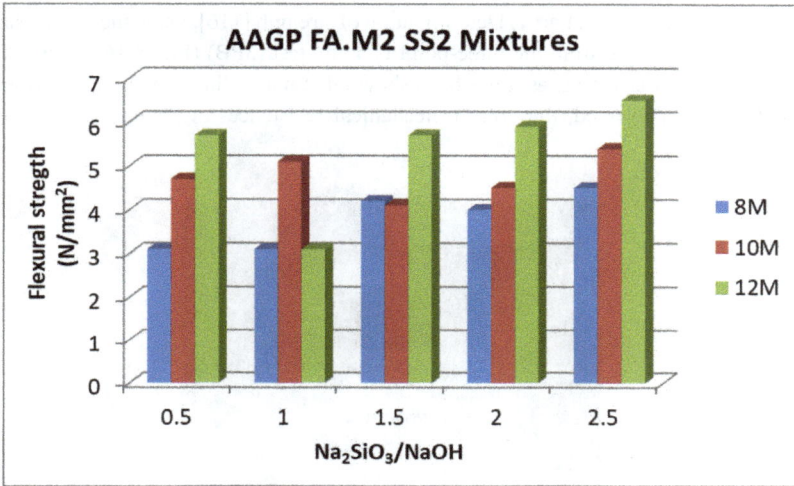

Figure 4.14. AAGP FA.M2 SS2 – 3PB.

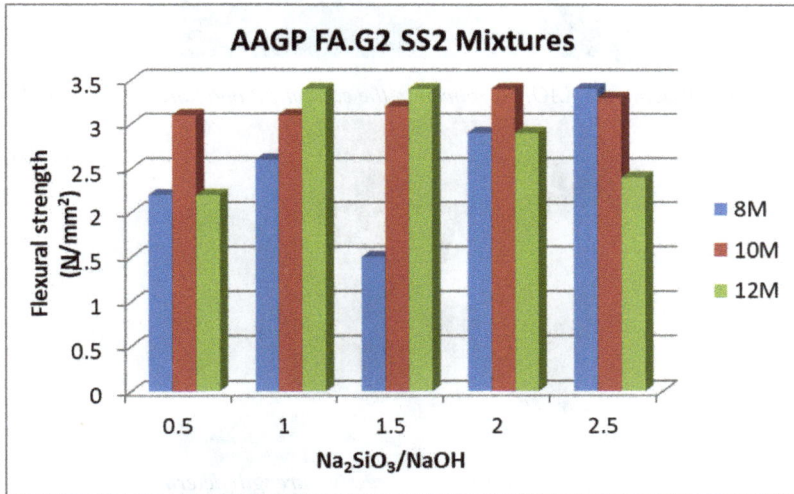

Figure 4.15. AAGP FA.G2 SS2 – 3PB.

Compressive strength of test specimens was determined in accordance with EN 196-1 Methods of testing cement. Part 1: Determination of strength [116], using the half prismatic test specimens resulting from the three-point bending test (3PB) (Fig. 4.16). The test age of the specimens was 7 days, as it has been shown that due to the heat treatment to which the mixtures are subjected, they reach mechanical resistances at young ages [39] (Fig. 4.17).

a) . Placing the AAGP specimen in the equipment and failure under load

b) Test specimens after compressive strength determination

Figure 4.16. Compressive strength testing of the samples.

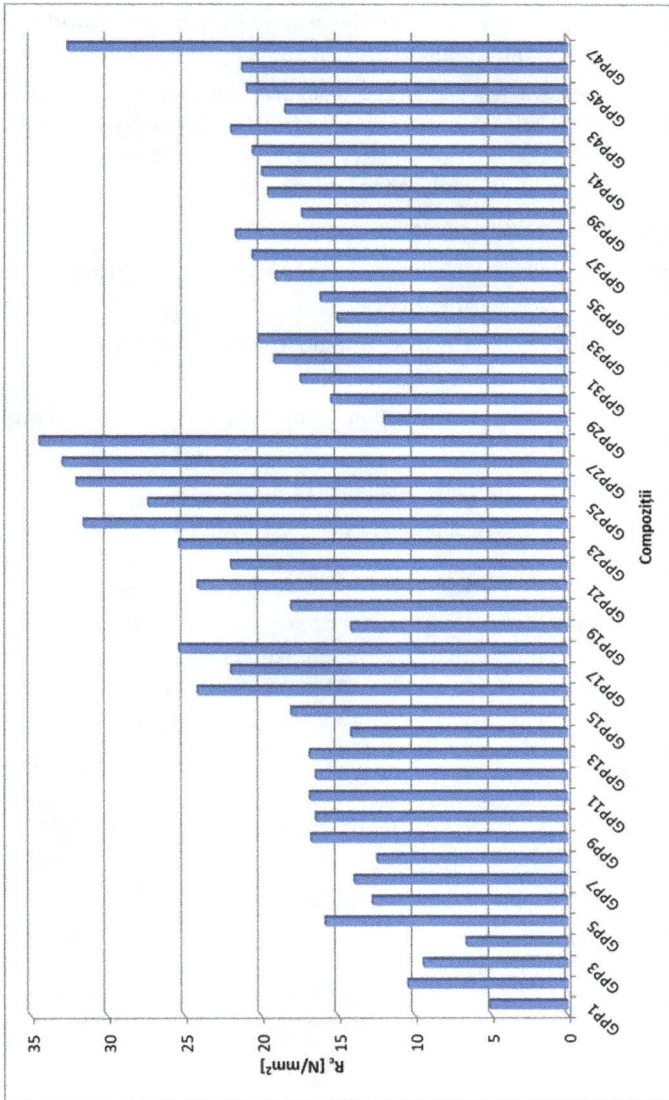

Figure 4.17. AAGP compressive strength.

4.4.3. Technological and mix-design influences on the performance of alkali-activated geopolymer binder

Compressive strength was chosen as the main parameter for the overall comparative assessment, therefore all mixtures were tested for compressive strength results; subsequently, certain additional parameters that influence the compressive strength of AAGP were investigated for the relevant mixtures:

- influence of age on the compressive strength;
- influence of the alkaline activator to fly ash ratio (AA/FA) on the compressive strength;
- influence of the alkaline activators ratio (Na_2SiO_3/NaOH) on the compressive strength;
- influence of the molar concentration of the NaOH solution on the compressive strength;
- influence of fly ash type on the compressive strength;
- influence of fly ash physical properties on the compressive strength.

4.4.3.1. Influence of age on the compressive strength

In order to establish the *influence of age on compressive strength*, the samples produced using fly ash G1 and M1, respectively, were selected, for which the alkaline activator was prepared with sodium hydroxide solution 8M and sodium silicate solution SS1 and SS2 (GPP1-GPP8). The compressive strength tests were carried out at 2, 7 and 28 days. The experimental results are shown in Fig. 4.18.

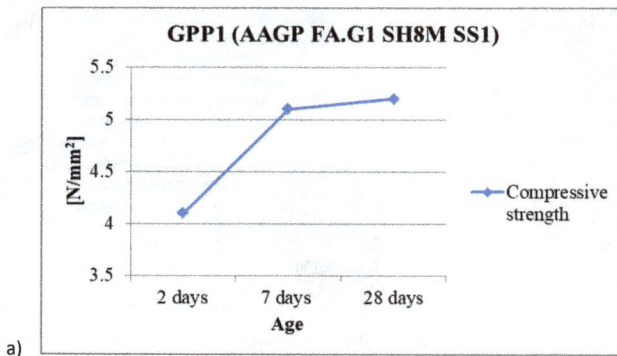

a)

b)

c)

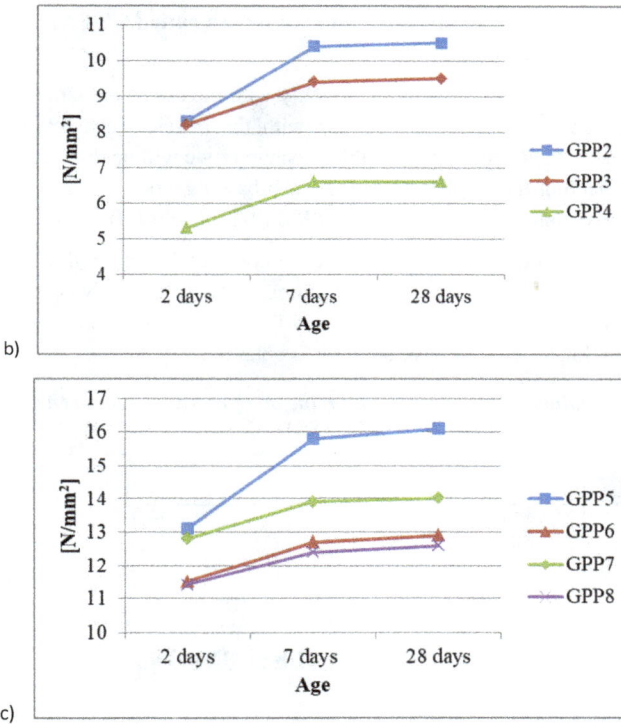

Figure 4.18. Influence of age on the compressive strength: a) Mixture GPP1; b)
Mixtures GPP2-GPP4; c) Mixtures GPP5-GPP8.

The compressive strength results, obtained at specific ages of 2, 7 and 28 days, respectively, indicate without exception that compressive strength at the age of 7 days reached values close to those at the age of 28 days (the reference age of the OPC concrete, at which the representative value is considered reached). Based on this analysis, for future evaluations, compressive strength at the age of 7 days will be considered a relevant indicator, equivalent to that of 28 days, but much more accessible in terms of compositional analysis times. The obtained conclusion is also confirmed by the literature [70].

4.4.3.2. Influence of the alkaline activator to fly ash ratio (AA/FA) on the compressive strength

This parameter has a close connection not only with the use of the same types of materials for obtaining geopolymer paste, but also with the physical characteristics of the raw material used for their production. In this subchapter we will analyze the influence of this parameter, selecting the alkali-activated fly ash-based geopolymer paste, produced using the same source material (AAGP FA.R SH 8M SS2) (Table 4.9).

It can be seen, according to Fig. 4.19 that the compressive strength of AAGP increased with the decrease of the ratio between the amount of alkaline activator and the amount of fly ash in mixtures. As in the case of ordinary Portland cement concrete, for a lower water / cement ratio, higher compressive strength is obtained.

Table 4.9.   Influence of AA/FA ratio on the compressive strength (AAGP FA.R SH 8M SS2).

Mixture	AA/FA	Compressive strength [N/mm²]
8M	26.23	73.77
10M	31.37	68.63
12M	36.09	63.91

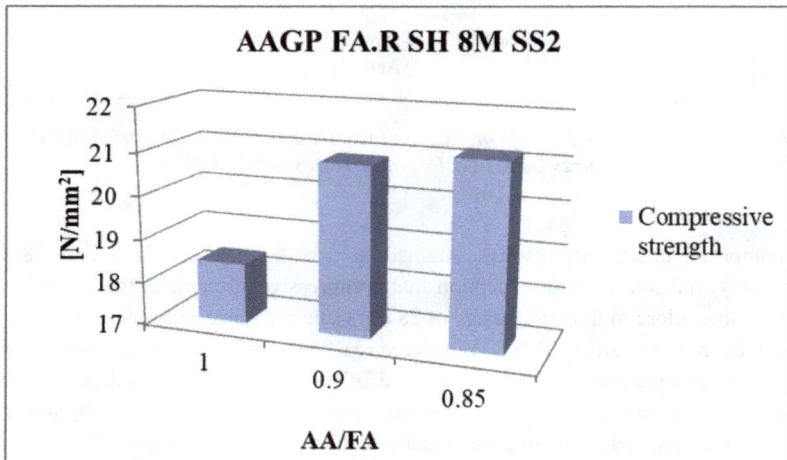

Figure 4.19. Influence of AA/FA ratio on the compressive strength (AAGP FA.R SH 8M SS2).

4.4.3.3. Influence of the alkaline activators ratio (Na₂SiO₃/NaOH) on the compressive strength

In order to establish the influence of the alkaline activators ratio on the compressive strength of the alkali-activated fly ash-based geopolymer paste the next mixtures have been selected:

- AAGP FA.M2 SH 8M SS2 (GPP14 – GPP18);

- AAGP FA.M2 SH 10M SS2 (GPP19 – GPP23);

- AAGP FA.M2 SH 12M SS2 (GPP24 – GPP28);

- AAGP FA.G2 SH 8M SS2 (GPP29 – GPP33);

- AAGP FA.G2 SH 10M SS2 (GPP34 – GPP38);

- AAGP FA.G2 SH 12M SS2 (GPP39 – GPP43);

The results obtained (Fig. 4.20 and Fig. 4.21) regarding the influence of the sodium silicate solution to sodium hydroxide solution ratio (Na_2SiO_3/NaOH) highlight the importance of this parameter regarding the final results of the compressive strength of the alkali-activated fly ash-based geopolymer paste. The evolution of the compressive strength values for the analyzed compositions, at the age of 7 days, is different and as this ratio increased, the values of mechanical properties of the material also increased. The only exception were some of the mixtures - AAGP FA.M2 SH 8M SS2 – 2,00 / AAGP FA.M2 SH 10M SS2 – 1,50 and AAGP FA.M2 SH 12M SS2 – 1,00 – where a slight decrease in the compressive strength values was observed.

For all AAGP mixtures that were analyzed, it can be seen that for a Na_2SiO_3/NaOH ratio of 2.5 the highest values for the compressive strength are obtained: 34.4 MPa for mixtures in which M2 fly ash was used, respectively 21.9 MPa for mixtures in which G2 fly ash was used.

It can be seen that the increase rate of the compressive strength is not constant, being different for each mixture. Thus, for the mixtures in which M2 fly ash was used, the growth percentages of the compressive strength, as the ratio of Na_2SiO_3/NaOH increased were: 79%, for NaOH 8M, 31% for NaOH 10M and 10% for NaOH 12M, respectively. Regarding the mixtures in which G2 fly ash was used, the growth percentages of the compressive strength were 68% for NaOH 8M, 44% for NaOH 10M and 26% for NaOH 12M, respectively.

The increase in the compressive strength values of the samples mainly depends on how the geopolymerization reaction is produced. A higher Na_2SiO_3/NaOH solution ratio adds to the geopolymer matrix more SiO_2 species, therefore more Si-O-Si bonds are formed, creating a stronger material.

Due to the interdependence between the many oxides that form the geopolymer matrix, it can be said that the influence of the alkaline activators ratio have on the material is very important, and certain decreases in the compressive strength are due to an incomplete geopolymerization process [13].

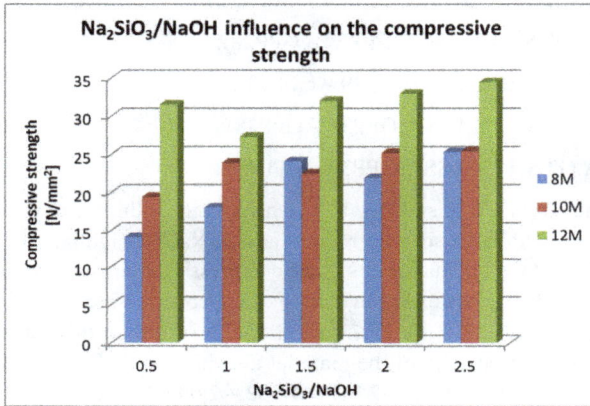

Figure 4.20. $Na_2SiO_3/NaOH$ influence on the compressive strength (AAGP FA.M2 SS2).

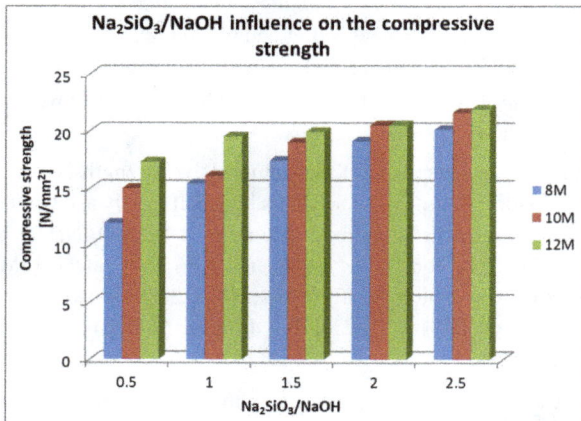

Figure 4.21. $Na_2SiO_3/NaOH$ influence on the compressive strength (AAGP FA.G SS2).

4.4.3.4. Influence of the molar concentration of the NaOH solution on the compressive strength

The influence of the molar concentration of the NaOH solution on the compressive strength of the alkali-activated fly ash-based geopolymer paste was evaluated on relevant mixtures, which have as common parameters: 1) the raw material (source and batch); 2) the Na_2SiO_3/NaOH solution ratio. These were grouped in the following categories (Fig. 4.22) and the results are presented in Fig. 4.23 a) –e):

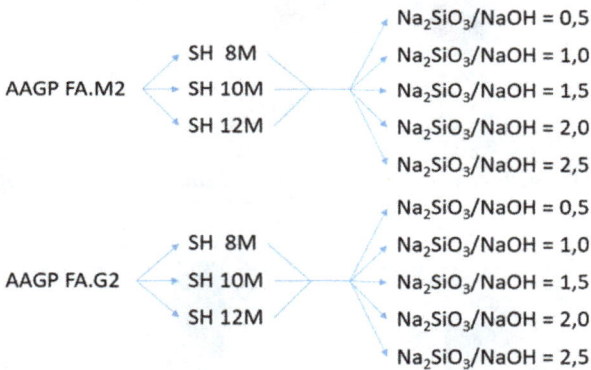

AAGP FA.M2

- SH 8M
- SH 10M
- SH 12M

- Na_2SiO_3/NaOH = 0,5
- Na_2SiO_3/NaOH = 1,0
- Na_2SiO_3/NaOH = 1,5
- Na_2SiO_3/NaOH = 2,0
- Na_2SiO_3/NaOH = 2,5

AAGP FA.G2

- SH 8M
- SH 10M
- SH 12M

- Na_2SiO_3/NaOH = 0,5
- Na_2SiO_3/NaOH = 1,0
- Na_2SiO_3/NaOH = 1,5
- Na_2SiO_3/NaOH = 2,0
- Na_2SiO_3/NaOH = 2,5

Figure 4.22. Analysis of the influence of molar concentration of NaOH solution.

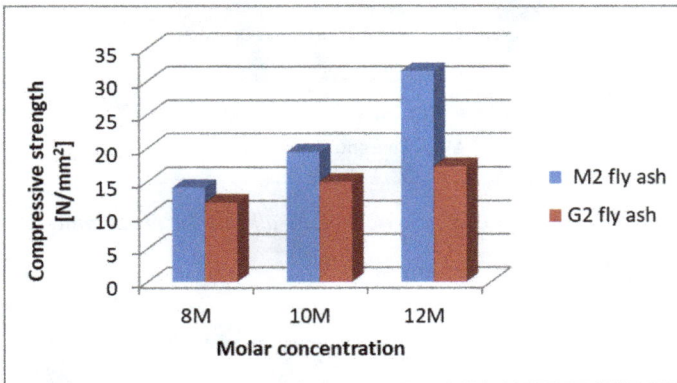

a) NaOH concentration influence – alkaline activator ratio 0.5

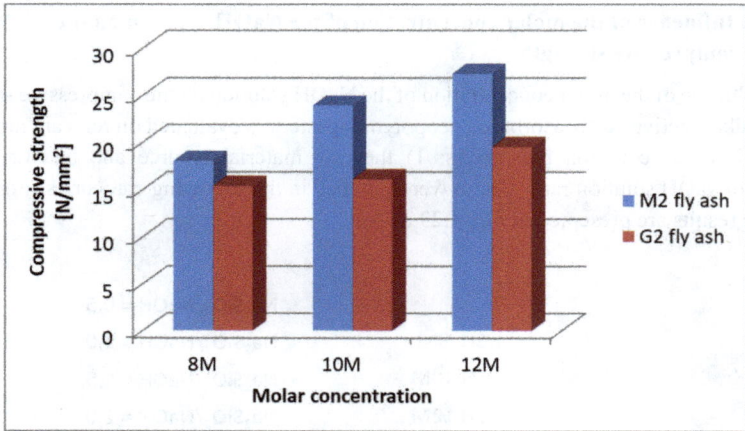

b) NaOH concentration influence – alkaline activator ratio 1.0

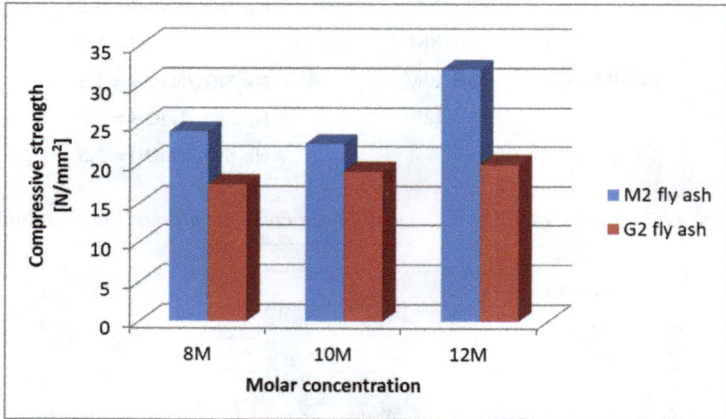

c) NaOH concentration influence – alkaline activator ratio 1.5

Materials Research Forum LLC
https://doi.org/10.21741/9781644901533

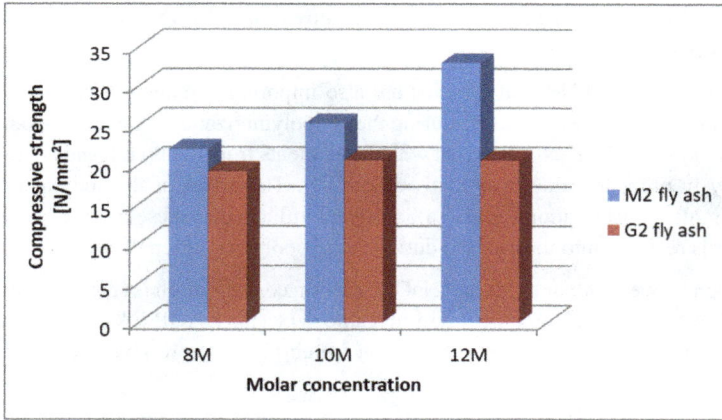

d) NaOH concentration influence – alkaline activator ratio 2.0

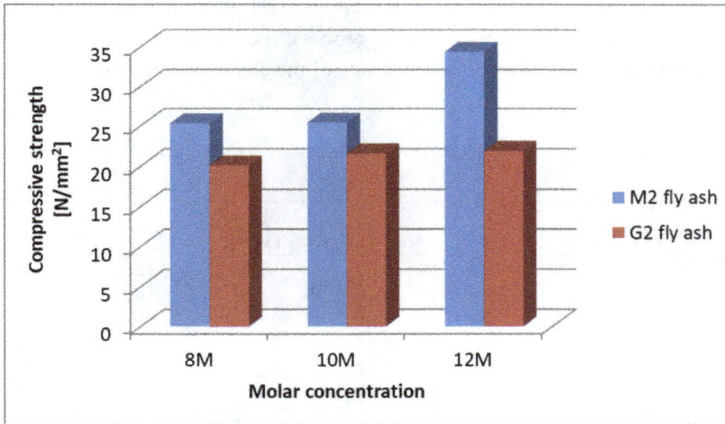

e) NaOH concentration influence – alkaline activator ratio 2.5

Figure 4.23. NaOH concentration influence on the compressive strength of AAGP samples.

4.4.3.5. Interdependence between molar concentration of NaOH solution and water in the mixture

Molar Na_2O/SiO_2 and H_2O/Na_2O ratios are also important parameters in terms of water released in the geopolymer mixture during the geopolymerization process. The constituent elements of the alkaline activator react with the elements from the fly ash and release water, necessary in order to achieve optimal workability. The effect of the sodium hydroxide solution molar concentration has also a significant influence on these parameters and, thus, the water is released into the mixture during the geopolymerization process.

Correlation between molar ratios and NaOH solution concentration is summarized in Table 4.10 (for AAGP FA.M2 mixtures) and in Table 4.11 (for AAGP FA.G2 mixtures). The reference parameter of the mixture evaluation remains the compressive strength at the age of 7 days.

By keeping constant alkaline activator to fly ash ratio (AA / FA) and constant alkaline activators ratio ($Na_2SiO_3/NaOH$) it was determined that molar Na_2O/SiO_2 ratio increased with the increase of the molar concentration of the NaOH solution. This molar ratio played an important role in the geopolymerization process, increasing the concentration of the NaOH solution lead to an higher dissolution rate of the elements in the source material (fly ash).

Table 4.10. NaOH solution concentration influence on the molar ratios (AAGP FA.M2 mixtures).

Mixture	$Na_2SiO_3/$ NaOH	NaOH Conc	Molar ratio			R_{cmed} N/mm²
			Na_2O/SiO_2	SiO_2/Al_2O_3	H_2O/Na_2O	
GPP14		8M	0.16	3.80	12.742	14.1
GPP19	0.5	10M	0.18	3.80	10.793	19.4
GPP24		12M	0.20	3.80	9.376	31.5
GPP15		8M	0.15	3.96	12.734	18.0
GPP20	1.0	10M	0.16	3.96	11.150	23.9
GPP25		12M	0.18	3.96	9.945	27.3
GPP16		8M	0.14	4.06	12.729	24.1
GPP21	1.5	10M	0.15	4.06	11.394	22.5
GPP26		12M	0.16	4.06	10.349	32.0
GPP17		8M	0.13	4.13	12.725	21.9
GPP22	2.0	10M	0.14	4.13	11.572	25.2
GPP27		12M	0.15	4.13	10.648	32.9
GPP18		8M	0.13	4.17	12.722	25.3
GPP23	2.5	10M	0.14	4.17	11.707	25.4
GPP28		12M	0.14	4.17	10.881	34.4

Materials Research Forum LLC
https://doi.org/10.21741/9781644901533

Table 4.11. NaOH solution concentration influence on the molar ratios (AAGP FA.G2 mixtures).

Mixture	Na_2SiO_3/ NaOH	NaOH Conc	Molar ratio			R_{cmed} N/mm^2
			Na_2O/SiO_2	SiO_2/Al_2O_3	H_2O/Na_2O	
GPP29		8M	0.28	4.95	13.177	11.9
GPP34	0.5	10M	0.32	4.95	11.116	15.0
GPP39		12M	0.36	4.95	9.628	17.3
GPP30		8M	0.25	5.34	13.194	15.4
GPP35	1.0	10M	0.28	5.34	11.512	16.1
GPP40		12M	0.30	5.34	10.241	19.5
GPP31		8M	0.23	5.57	13.206	17.4
GPP36	1.5	10M	0.25	5.57	11.785	19.0
GPP41		12M	0.27	5.57	10.678	19.9
GPP32		8M	0.22	5.72	13.214	19.1
GPP37	2.0	10M	0.23	5.72	11.983	20.5
GPP42		12M	0.25	5.72	11.003	20.5
GPP33		8M	0.21	5.83	13.220	20.1
GPP38	2.5	10M	0.22	5.83	12.136	21.6
GPP43		12M	0.24	5.83	11.257	21.9

The workability of the mixtures was also assessed considering the variation of the molar H_2O/Na_2O ratio. Thus, it can be seen a clear interdependence between the variation of the workability and compressive strength of the mixtures, and the molar H_2O/Na_2O ratio: direct proportionality – in case of workability (the lower the ratio, the lower the workability) and inversely proportional – in case of compressive strength (the lower the ratio, the higher the compressive strength).

Based on the results obtained and presented in this chapter, it can be said that for the development of alkali-activated fly ash-based geopolymer materials, the influence of the sodium hydroxide solution concentration is a very important parameter, not only in the geopolymerization process, but also in the mechanical properties of the material.

The solubility of the aluminosilicate material (fly ash) increases as the concentration of the NaOH solution increases. At the same time, a higher concentration of the solution accelerates the formation of the aluminum-silicate gel, due to the high alkaline environment, increasing the amount of hydroxide ions (OH⁻) in the mixture.

In the case of AAGP, it could be observed that as the NaOH solution concentration increased, the compressive strength of the mixtures increased. The growth percentages of compressive strength are different, and are shown in Fig. 4.24 and Fig. 4.25.

Also, the interdependence between the molar concentration of the sodium hydroxide solution and the molar Na_2O/SiO_2 and H_2O/Na_2O ratios revealed that as the concentration of the solution increased, a decrease in the workability of the compositions could be observed, by decreasing the amount of mixing water, but that also led to an increase in the mechanical strength of the material.

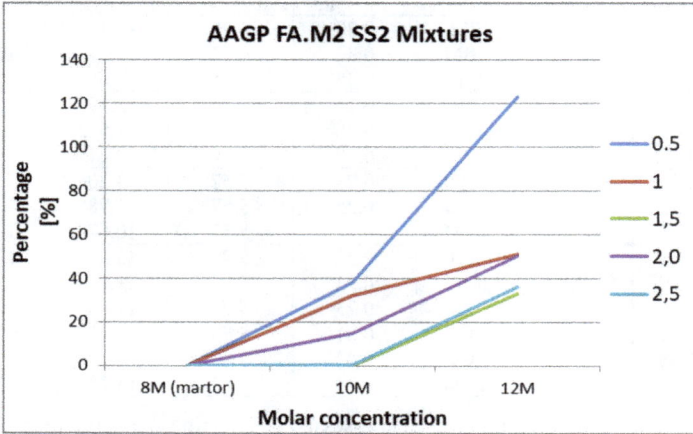

Figure 4.24. Compressive strength growth based on the NaOH molar concentration – AAGP FA.M2 SS2.

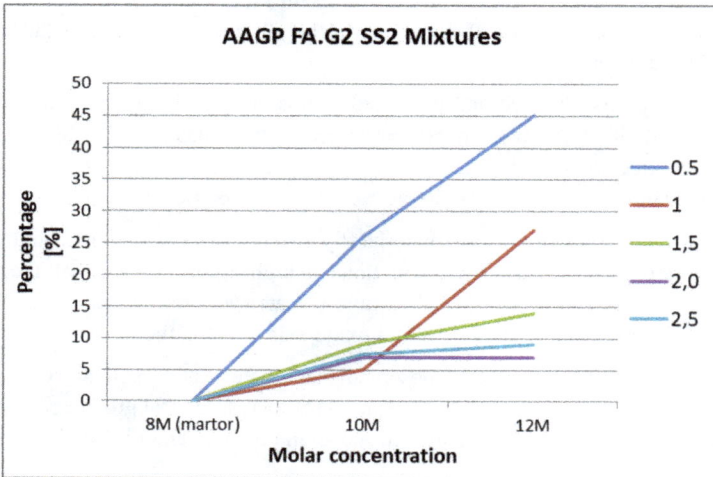

Figure 4.25. Compressive strength growth based on the NaOH molar concentration – AAGP FA.G2 SS2.

Materials Research Forum LLC
https://doi.org/10.21741/9781644901533

Based on the results presented above, it can be noted that for AAGP FA.M2 SS2 mixtures, compressive strength increased with the increase of the NaOH solution concentration from 8M to 12M in the range 36-51%, except for mixtures with Na_2SiO_3 / NaOH = 0.5 at which the percentage of increase of compressive strength was 123%. In the case of AAGP FA.G2 SS2 mixtures, the percentage of increase in compressive strength was in the range 7-45%.

4.4.3.6. Fly ash type influence on the compressive strength

The chemical composition and physical characteristics of fly ash are determined by the type of coal burned and the type of processing to which it is subjected. In order to determine the influence that the type of fly ash, namely its chemical composition, has on the compressive strength of the geopolymer binders, the following mixtures were selected which will be analyzed in Fig. 4.26, Fig. 4.27 and Fig. 4.28:

- AAGP FA.M2 and FA.G2 – SH 8M SS2 (Fig. 4.26);

- AAGP FA.M2 and FA.G2 – SH 10M SS2 (Fig. 4.27);

- AAGP FA.M2 and FA.G2 – SH 12M SS2 (Fig. 4.28).

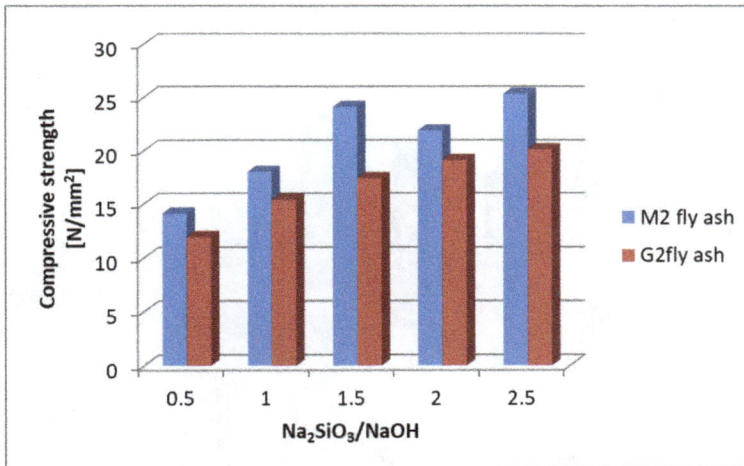

Figure 4.26. Influence of fly ash type on the compressive strength (AAGP FA.M2 and FA.G2 - SH 8M SS2).

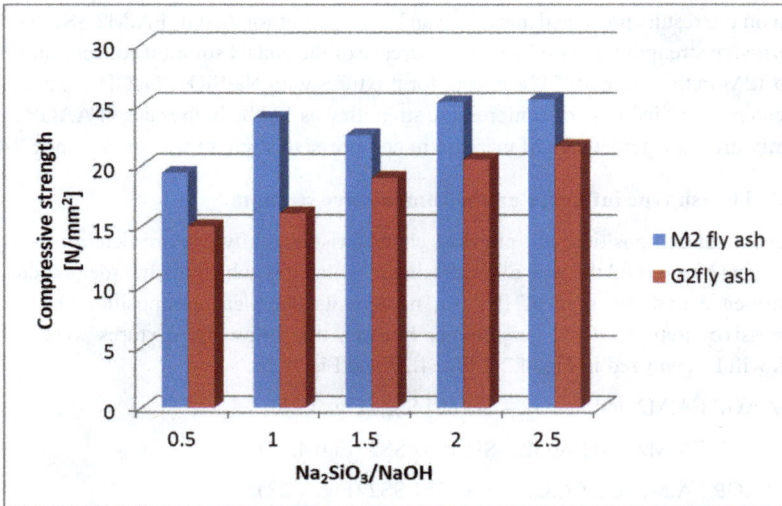

Figure 4.27. Influence of fly ash type on the compressive strength (AAGP FA.M2 and FA.G2 - SH 10M SS2).

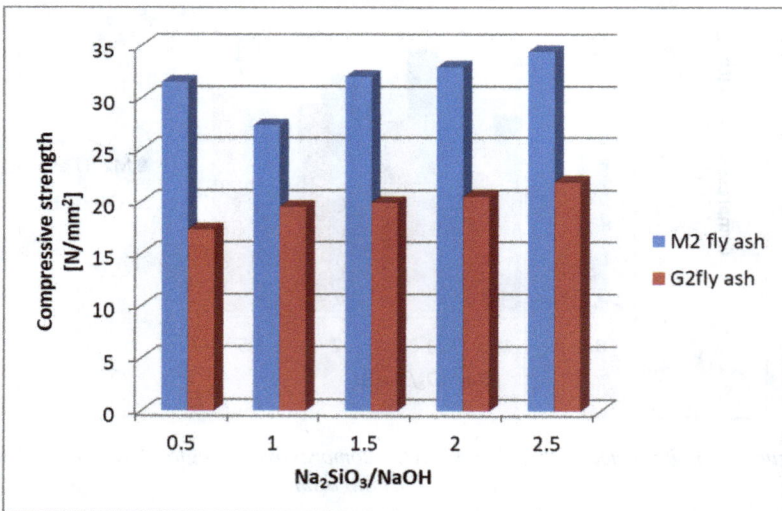

Figure 4.28. Influence of fly ash type on the compressive strength (AAGP FA.M2 and FA.G2 - SH 12M SS2).

Based on the results presented above it can be seen that in terms of compressive strength of AAGP, the chemical composition and physical characteristics of fly ash also play an important role. By preserving the molar concentration parameters of the NaOH solution and the Na_2SiO_3 / NaOH ratio constant, the importance of the chemical composition of the fly ash and its physical characteristics could be demonstrated.

It can be seen that for AAGP mixtures developed with M2 fly ash, the values of the compressive strength were superior to those in which G2 fly ash was used. Higher values of compressive strength of the mixtures produced with M2 fly ash are due to both chemical and physical characteristics of it (Table 4.12).

Table 4.12. Fly ash characteristics.

Chemical composition		
Oxides	Fly ash	
	G2	M2
SiO_2	54.32 %	53.61 %
Al_2O_3	22.04 %	26.16 %
Fe_2O_3	9.02 %	7.58 %
CaO	5.85 %	2.42 %
Loss on Ignition	3.69 %	3.57 %
Physical characteristics		
$R_{0,045}$ Fineness	51.60 %	37.60 %

It can be seen from Table 4.12 that, from a chemical point of view, the amount of the metal oxides, which have a direct relation to the geopolymerization process, and which, according to the ASTM (2019) must satisfy the conditions for the development of AAGM, are far superior for fly ash, M2: $SiO_2 + Al_2O_3 + Fe_2O_3 = 87.35$ %, respectively, 85.38 %, for the fly ash G2. It can also be seen that from the point of view of CaO content, M2 fly ash is superior to M2 fly ash, with a lower content of calcium oxide in the composition.

In terms of physical characteristics, M2 fly ash is also superior to G2 fly ash, in terms of fineness determined on the 0.045 mm sieve, and a finer fly ash leads to a stronger geopolymerization reaction, as a result, to higher compressive strength.

4.4.3.7. Influence of fly ash physical properties on the compressive strength

The chemical composition of the source material for the production of geopolymers plays an important role in the properties of the obtained material, but the physical characteristics also have an important role in the geopolymerization process and, later, in the development of the compressive strength. In this subchapter, results obtained on geopolymer paste mixtures analyzing the fineness parameter $R_{0.045}$ will be presented.

In order to establish the influence of this parameter on the compressive strength of the alkali-activated geopolymer paste, but also the ratio between the alkaline liquid and the fly plant ash (AA/FA), AAGP FA.R. SH 8M mixture were selected. Results are presented in Table 4.13 and Fig. 4.29.

Table 4.13. $R_{0.045}$ influence on the compressive strength.

Mixture	Na2SiO3/NaOH	AA/FA	$R_{0.045}$ %	R_{cmed} N/mm^2
GPP44		1.00		18.4
GPP45	1.00	0.90	31.40	20.9
GPP46		0.85		21.2
GPP47		0.75	0,00	32.6

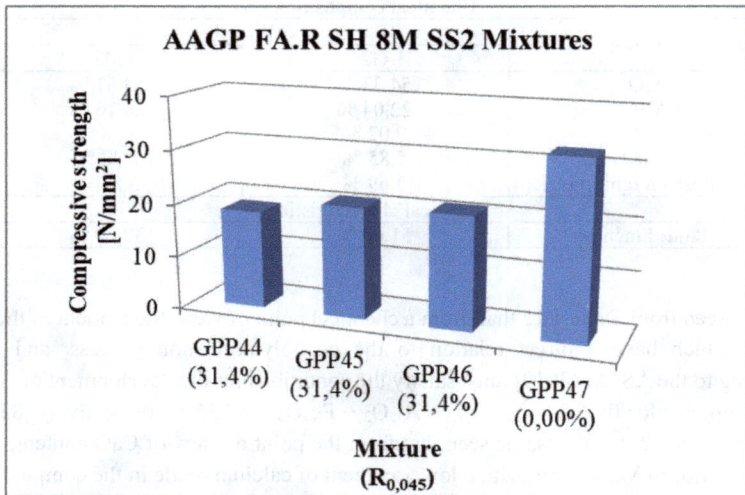

Figure 4.29. $R_{0.045}$ influence on the compressive strength.

From Table 4.13 and Fig. 4.29, it can be seen that the fineness of the fly ash also plays a very important role in the development of compressive strength. For fly ash on which no processing was carried out, the ratio between the alkaline liquid and the amount of fly ash could be decreased to 0.85, this generating the highest values of compressive strength. For values below this ratio, the geopolymerization reaction could not be achieved.

Regarding the fly ash on which grinding and sieving processes were performed on the characteristic sieve (0.045 mm) it can be observed a decrease of this ratio, a smaller amount of activator being needed to achieve the geopolymerization reaction. At the same time, by processing it, higher compressive strength values were obtained because the rate of dissolution of the particles was much higher, generating a much denser geopolymer matrix.

4.4.4. Microscopic evaluation of alkali-activated geopolymer paste mixtures

One of the most important microstructural parameters of alkali-activated geopolymer materials is porosity, but little progress has been made in terms of controlling or adapting mixtures to reduce this phenomenon.

Reducing porosity is essential for the physical-mechanical performances of the geopolymer materials. The considerable amount of water eliminated from the alkaline liquid during the geopolymerization reaction leads to the involuntary formation of mesopores. This is likely due to the fact that the porosity of the geopolymer material tends to be made up from both the mesopores resulting from the precipitation of the constituent materials, as well as from the resulting gaps between the matrix and the aggregates.

By microscopic analysis on test specimens for compressive strength (Fig. 4.30), both the porosity of the geopolymer matrix and the voids appeared in the microstructure of the material were observed.

Regarding certain alkali-activated geopolymer paste mixtures, it was observed the appearance of a white efflorescence, occurring during the conditioning period of samples under laboratory conditions (Fig. 4.31).

Efflorescence (due to the formation of white carbonate or bicarbonate crystals) is a known problem in alkaline-activated binders with too high concentration of sodium hydroxide solution, when excess alkali reacts with CO_2.

When using systems consisting of an alkaline liquid based on solutions with Na in the composition, consideration should be given to minimizing the amount of NaOH solution and using Na_2SiO_3/NaOH ratios that are convenient for generating the geopolymerization reaction.

It is also recommended to reduce the amount of water released from the system in order not to affect the mechanical properties of the geopolymer through the establishment of the migration of the ions to the surface of the hardened mixture.

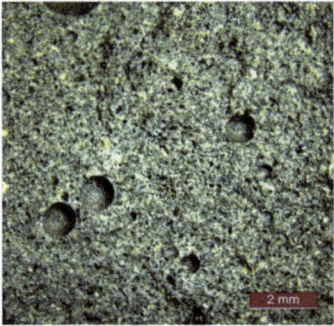

GPP29

GPP30

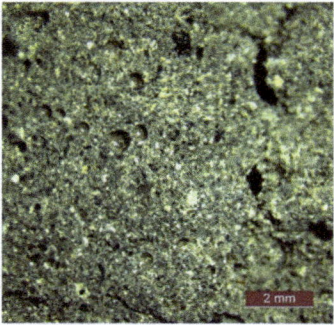

GPP31

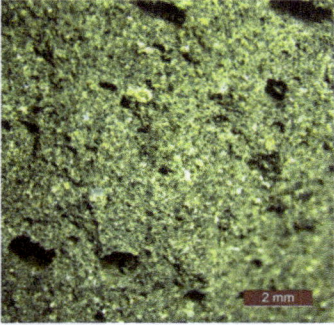

GPP32

GPP33

GPP34

Materials Research Forum LLC
https://doi.org/10.21741/9781644901533

GPP35

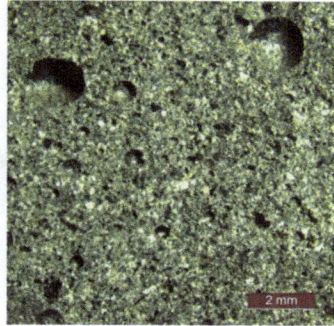

GPP36

GPP37

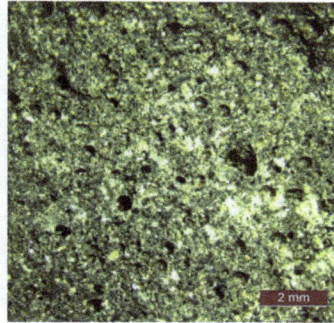

GPP38

GPP39

GPP40

GPP41

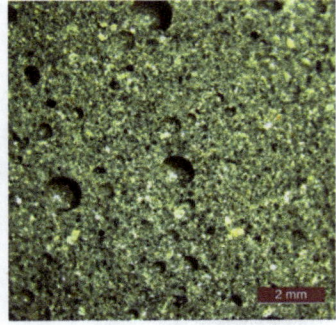

GPP42

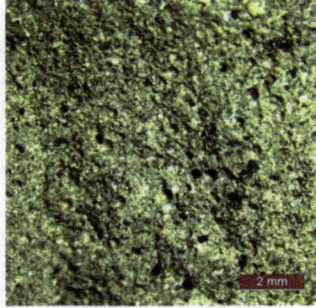

GPP43

Figure 4.30. Geopolymer matrix microscopic aspects.

(1.0x)   *(5.0x)*

(8.0x)

Figure 4.31. Na₂CO₃ crystals.

Chapter 5

Development of Alkali-Activated Fly Ash-Based Geopolymer Concrete

Preliminary alkali-activated geopolymer concrete mixtures were developed based on studying the literature and starting from a rigorous selection of source materials that will be used to produce this type of material. By choosing relevant research in the literature, the preliminary mixtures were produced based on the variation of the parameters influencing them, from the point of view of the specific ratios, as well as their chemical composition. The evolution of the geopolymer concrete mix design was developed based on the observations collected in the experimental research and initial practical assessments, the fresh state appearance and the performance of the materials at a certain time.

The sequence and duration of mixing operations was initially established, for the alkali-activated geopolymer paste mixtures and stabilized unitary within the first preliminary mixtures; thus, it was identified the need to extend the period of preparation of the alkaline activator and increase the time of its addition over the fly ash, in order to increase the efficiency of the geopolymerization reaction.

By modifying the variables and ratios of materials used in the production of geopolymer concrete (alkaline liquid to fly ash ratio, molar concentration of the NaOH solution, $Na_2SiO_3/NaOH$ ratio), several necessary data were obtained on the parameters that influence this type of material and its possible directions of applicability.

5.1. Raw materials characteristics

The principles developed in the previous chapter on alkali-activated geopolymer paste, were used for the preliminary alkali-activated fly ash-based geopolymer concrete mixtures. The materials used for their production were:

- Fly ash (FA) – G2 and M2;

- Sodium hydroxide solution (NaOH) – flakes, 98% purity – SH1 and SH2;

- Sodium hydroxide solution (NaOH) – pearls, 99% purity – SH3;

- Sodium silicate solution (Na_2SiO_3) – SS2;

- Natural aggregates, granular class 0/4 mm (S) and 4/8 mm (CA), (sand and coarse aggregates).

5.2. Alkali-activated fly ash-based geopolymer technology

As in the case of alkali-activated geopolymer paste production presented in the previous chapter, a paddle mixer, with 4.75 l capacity, a rotational movement of 140/280 rot/min and 61.5/123 rot/min revolution, was used to produce alkali-activated fly ash-based geopolymer concrete. Mixing technology, specific to low volume experimental casting (maximum 0.8-1.0 l) included the following sequences:

- Conditioning of the raw material (fly ash and aggregates) at $(20\pm2)°C$;

- Preparation of the alkaline activator by combining the Na_2SiO_3 and NaOH solutions 24 hours prior;

- The sand (0/4 mm) and the coarse aggregates (4/8 mm) were placed in the mixing container and mixed homogenous, at low speed, for 30 seconds;

- The necessary amount of fly ash was then added and the raw materials (fly ash + aggregates) were mixed together for 30 seconds;

- The alkaline activator was added during 90 seconds over the homogenous mixture;

- After adding the total alkaline activator quantity, the mixing was continued at low speed for 3 minutes during which, the workability of the mixture has been observed;

- *for some mixtures, when low workability was observed, superplasticizer was added, in order to evaluate its effect on the mixtures;

- The mixtures were then placed in 40 mm x 40 mm x 160 mm molds and heat cured at 70°C for 24 hours. A glass film was placed on top of every mold in order to prevent excessive water release from the mixtures;

- After demolding, the specimens were stored in the climatic chamber at the temperature T $(20\pm1)°C$ and relative humidity RH $(60\pm5)\%$.

The mechanical performances were tested at 7 days of age.

To achieve a homogeneous mixture, as in the case of alkali-activated geopolymer paste, the same factors that influence the geopolymerization were taken into account: alkaline activator to fly ash ratio (AA/AF.), the ratio of alkaline activators ratio (SS/SH). Furthermore, in order to optimize the mixtures, the percentage occupied by aggregates in the mass of the mixture in order to achieve satisfactory workability should be considered.

5.3. Alkali-activated fly ash-based geopolymer concrete mixtures

A number of five preliminary alkali-activated geopolymer concrete mixtures were produced (AAGC1–AAGC5), subsequently derived in a number of 24 secondary mixtures

(AAGC6-AAGC29) (Table 5.1), grouped according to the influence of the component materials and aimed at their comparison based on the results obtained for the flexural, respectively compressive strength, at early age (7 days).

Note: In order to simplify the presentation of the obtained results, al the alkali-activated fly ash-based geopolymer paste samples have been grouped in several categories, based on the primary raw materials that were used in their production (Eq. 5.1).

AAGC FA αx SH yM SSz   (5.1)

Where: *AAGC* – alkali-activated geopolymer concrete; *FA* – fly ash-based; *αx* – type of fly ash; *SH yM* – sodium hydroxide solution with y molarity; *SSz* – type of sodium silicate solution.

Table 5.1.   AAGC mix design ratios.

Mixture	Fly ash type	AA/ FA	Ratio to fly ash	Na_2SiO_3 / NaOH	SS	SH	NaOH conc.	Sand 0/4 mm	Agg. 4/8 mm
AAGC1*				1.0	0.5	0.5		0.76	0.51
AAGC2				1.0	0.5	0.5	8M SH1	0.57	0.57
AAGC3*	G2	1.0	1.0	1.0	0.5	0.5		0.5	0.5
AAGC4				1.0	0.5	0.5		0.5	0.5
AAGC5				2.0	0.67	0.33		0.5	0.5
AAGC6				0.5	0.17	0.33			
AAGC7				1.0	0.25	0.25	8M SH1		
AAGC8	G2	1.0	1.0	1.5	0.30	0.20		0.5	0.5
AAGC9				2.0	0.33	0.17			
AAGC10				2.5	0.36	0.14			
AAGC11				0.5	0.17	0.33			
AAGC12				1.0	0.25	0.25	8M SH2		
AAGC13	G2	1.0	1.0	1.5	0.30	0.20		0.5	0.5
AAGC14				2.0	0.33	0.17			
AAGC15				2.5	0.36	0.14			

For AAGC1 and AAGC3 a superplasticizer admixture (0.2 proportion of fly ash, FA) was used in order to optimize the workability, observed as having an unsatisfactory evolution during mixing.

Mixture	Fly ash type	AA/ FA	Ratio to fly ash	Na$_2$SiO$_3$ / NaOH	SS	SH	NaOH conc.	Sand 0/4 mm	Agg. 4/8 mm
AAGC16				0.5	0.17	0.33			
AAGC17				1.0	0.25	0.25			
AAGC18	G2	1.0	1.0	1.5	0.30	0.20	10M SH2	0.5	0.5
AAGC19				2.0	0.33	0.17			
AAGC20				2.5	0.36	0.14			
AAGC21	M2	0.6	0.6	2.5	0.42	0.17	12M SH2	0.5	0.5
AAGC22				1.0	0.25	0.25			
AAGC23	G2	1.0	1.0	1.5	0.30	0.20	8M SH3	0.5	0.5
AAGC24				2.0	0.33	0.17			
AAGC25	G2	1.0	1.0	2.5	0.36	0.14	8M SH3	0.5	0.5
AAGC26				1.0	0.25	0.25			
AAGC27	G2	1.0	1.0	1.5	0.30	0.20	10M SH3	0.5	0.5
AAGC28				2.0	0.33	0.17			
AAGC29				2.5	0.36	0.14			

5.4. Alkali-activated fly ash-based geopolymer physical-mechanical properties

5.4.1. Fresh-state properties

Similar to the development process of the alkali-activated fly ash-based geopolymer paste (AAGP), an important aspect of the development of alkali-activated fly ash-based geopolymer concrete, is workability. When producing such mixtures, it is important to take into account the parameters of the compositional design and, therefore, to record at any time of mixing the behavior of fresh mixtures. Also, based on observations from the geopolymer paste tests, it was found that when the Na$_2$SiO$_3$/NaOH ratio increases, the workability of the geopolymer mixture also increases. The functionality of mixtures with NaOH 10M solution was considerably inferior to mixtures with NaOH 8M solution; thus the importance of the concentration of NaOH solution, similar to AAGP mixtures.

As with the production of alkali-activated geopolymer paste, for the for all the alkali-activated geopolymer concrete mixtures, a slight decrease in the workability was observed as the NaOH solution concentration increased from 8M to 10M. This happens mainly because, as the concentration of the solution increases, the amount of water released into the mixture is lower (NaOH 8M = 73.8% water / NaOH 10M = 68.6% water).

Although initially, mixtures appeared to be dry, unlikely to be molded, as shocks were applied for their compaction water started to be released in the mixture and workability was improved (Fig. 5.1).

For a small number of alkali-activated geopolymer concrete samples (AAGC6, AAGC11 and AAGC16) flash setting occurred. Based on the work of Black [118], this mainly happened both due to the increased quantity of NaOH solution in the mixture and its molar concentration, but also due to the chemical composition of the fly ash that was used to produce them. For this mixtures neither flexural strength, nor compressive strength were obtained. In order to establish if flash-setting is an isolated phenomenon, the mixtures and the conditions should be reconstituted. The evaluation of this phenomenon represents a desideratum of future research in the field of alkali-activated fly ash-based geopolymer concrete.

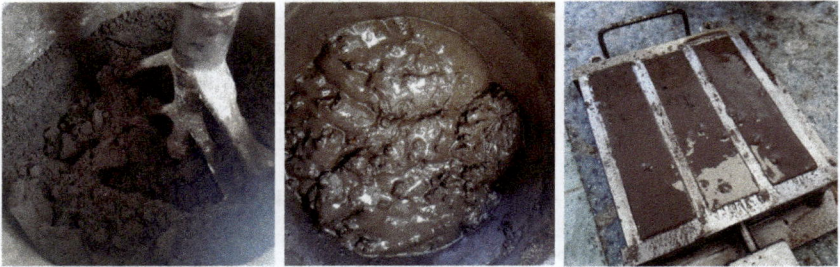

Figure 5.1.   AAGC fresh state properties.

5.4.2. Apparent density

The apparent density of the preliminary alkali-activated fly ash-based geopolymer concrete samples was determined in the same way as those for the geopolymer paste (Chapter 4), according to EN 12390-7 [117], at 7 days of age. Results are presented in Fig. 5.2.

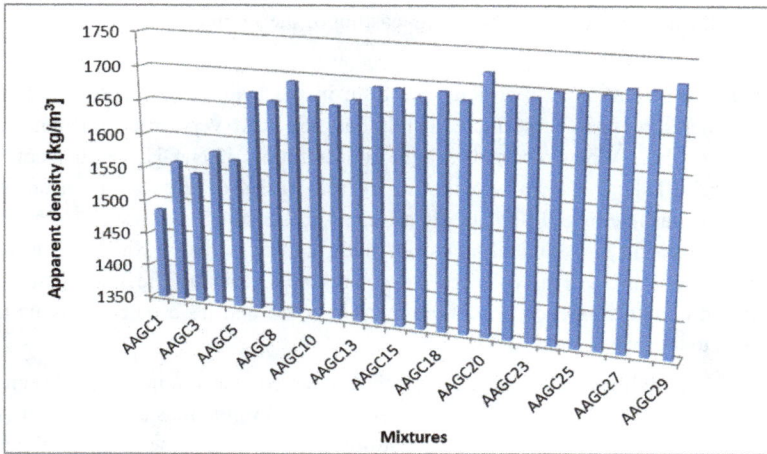

Figure 5.2.   AAGC apparent density.

5.4.3. Flash setting

The term "flash-setting" is attributed to geopolymer mixtures when, during mixing, an almost instant hardening of the mixture occurs, with very high heat generation. Upon completion of mixing, for a small number of alkali-activated geopolymer concrete mixtures, "flash-setting " occurred (Fig. 5.3 and Fig. 5.4). The mixtures for which this phenomenon occurred are the mixtures in which the $Na_2SiO_3/NaOH$ ratio was 0.5 and for the mixtures in which the G2 fly ash was to produce mixtures with a NaOH solution concentration of 12M.

Therefore, it can be noted that these mixtures have a large amount of alkaline activator, an average amount of water and a large amount of NaOH solution. This is an important observation, because if it happened in a practical application, with high volume mixtures, it could have significant consequences.

The samples that presented flash-setting presented the following characteristics:

- Flash-setting occurred during mixing or at maximum 2-3 minutes after the end of mixing;

- The temperature of the mixture increased quickly, which led to an instant geopolymerization reaction;

- After the appearance of flash-setting, casting of the samples was almost impossible, the samples breaking into large pieces.

This phenomenon has been reported very little in the available geopolymer materials literature; therefore it is very likely that it has not been very often encountered by researchers to date. When reported, a high concentration of NaOH was present in the mixture [15,119]. High calcium content in fly ash was also given as a reason for the occurrence of flash-setting [120]. The fly ash used in this study had low calcium content, therefore this reason is excluded. However, both high amounts of NaOH solution or a higher concentration were used in the current research, causing the flash-setting. It is believed that this phenomenon occurs at a certain point, and it's a potential issue that is meaningful to the users of the AAGC.

The current research regarding flash-setting that appeared, was not necessarily seen as an isolated phenomenon, but on the contrary, in some specific situations and characterized by some common elements, it is considered that this phenomenon presents a significant risk for the production and applications of AAGM. The exact reasons and proportions associated with the phenomenon are not yet sufficiently known, but will be the subject of future research, closely related to the areas of direct applicability and industrial production technology, in order to avoid similar situations

Figure 5.3.   Flash setting during mixing.

Figure 5.4.   Flash setting during casting.

5.4.4. Physical-mechanical properties of the alkali-activated geopolymer concrete mixtures

The tests regarding the 3PB bending tensile strength of AAGC compositions were carried out, as in the case of AAGP paste, in accordance with the methodology EN 196-1 [116], at the age of 7 days. The results obtained are shown in Fig. 5.5. Aspects of the behaviour of the material during the 3PB test are shown in Fig. 5.6 and Fig. 5.7. The compressive strength of AAGC samples was also determined similarly to AAGP samples, according to the methodology EN 196-1 [116]. The relevant age for the evaluation of the compressive strength of AAGC mixtures was 7 days. The results obtained are shown in Fig. 5.8. Aspects regarding the behavior of the material during the compressive strength test are shown in Fig. 5.9.

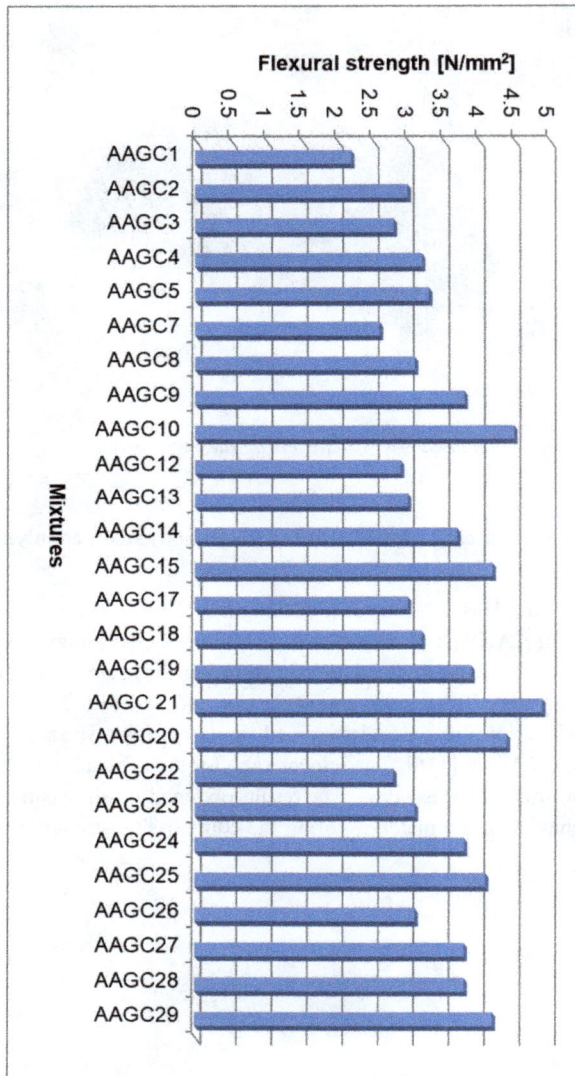

Figure 5.5.   AAGC – 3PB results.

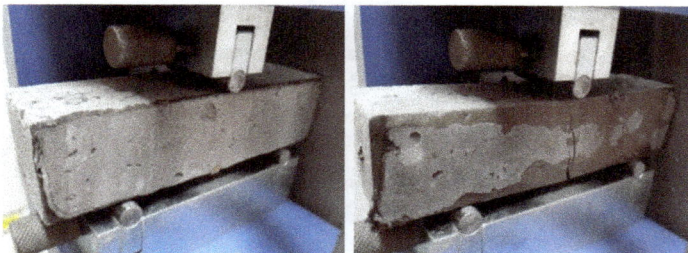

Figure 5.6.   AAGC – 3PB testing.

Figure 5.7.   AAGC – 3PB testing samples.

Materials Research Forum LLC
https://doi.org/10.21741/9781644901533

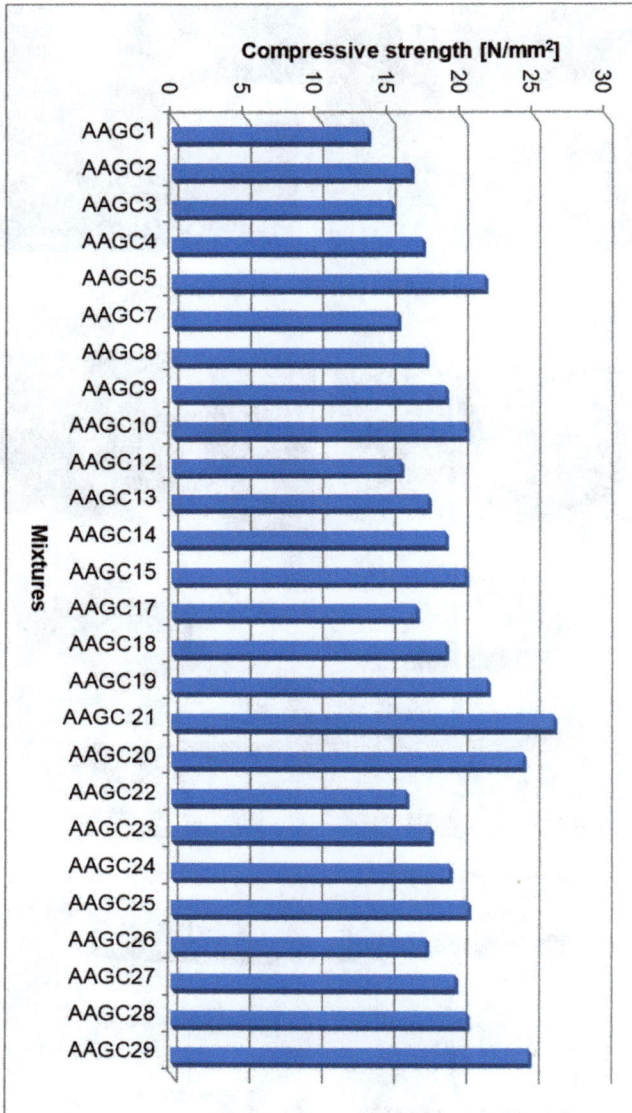

Figure 5.8.   AAGC – Compressive strength results.

Figure 5.9.   AAGC samples after the compressive strength test.

5.4.5. Microscopic evaluation

Since the binder of the alkali-activated geopolymer concrete is the alkali-activated geopolymer paste, it can be said that among its microstructural parameters, porosity is also important in the development of the material. There is little progress in the literature regarding the influence that this factor has on the mechanical strength and durability of this material. However, the involuntary formation of mesopores, due to the amount of water in the mixture, is a factor that should not be neglected. By microscopic evaluation (Fig. 5.10) observations were made on the porosity of the geopolymer matrix, but also on the voids within the mixtures.

Materials Research Forum LLC
https://doi.org/10.21741/9781644901533

The microstructure of concrete gives water the possibility to penetrate this type of material, therefore it is an essential factor related to the durability performance of the geopolymer concrete. The geopolymer concrete microstructure consists of dense, geopolymer paste, but also includes the interface between aggregate and it, as well as microcracks and air voids. They must be taken into account when dealing with the durability of this type of material.

At young ages, the microstructure of the alkali-activated geopolymer concrete is represented by a dense network of pores with a structure consisting of many voids and microcracks. This happens, perhaps, due to the fact that during the mixing the geopolymerization reaction was not complete and the water released during this process was in excess. By adding water or using alkaline activator molar ratios that release a large amount of water during the reaction lead to increased workability. If this happens, it can be seen that the structure of the geopolymer concrete has fewer pores and fewer voids (eg: AAGC8).

On the other hand, mixtures with smaller amount of water generate low workability of the mixture. The microstructure of these mixtures is a weak one, presenting many pores and many voids, but for these mixtures superior mechanical performances were achieved. This is completely different from the usual concrete microstructure, where for high porosity, lower mechanical characteristics are obtained.

Although alkali-activated geopolymer concrete has relatively many voids and microcracks, at relatively small ages it can have high mechanical performance. Independent of the variation in the chemical composition of the constituent materials, several factors play an essential role in the microstructure of geopolymer materials. Mixtures with a small amount of water, low workability, but also insufficient compaction of the mixture lead to a porous microstructure due to the amount of occlusal air. Studies have shown [121-124], that the amount of water used in the mixture, the chemical composition of the alkaline solution, the chemical composition of the fly ash, the aggregate-binder ratio, and methods of mixing have a very significant influence on the microstructure of the geopolymer concrete and for their final mechanical performances.

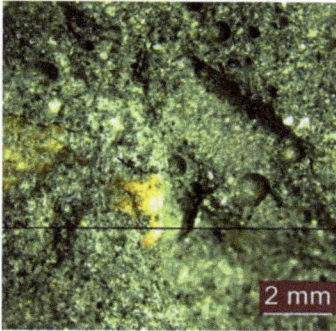

AAGC7

AAGC8

AAGC15

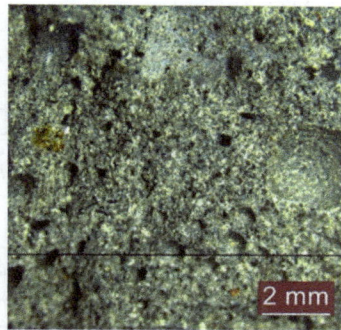

AAGC20

AAGC26

AAGC27

Figure 5.10. AAGC microscopic evaluation.

Materials Research Forum LLC
https://doi.org/10.21741/9781644901533

5.5. Technological and mix-design influences on the performance of alkali-activated geopolymer concrete

5.5.1. Influence of the molar concentration of the NaOH solution on the compressive strength

In order to study the influence that the NaOH solution concentration has on the compressive strength of the alkali-activated geopolymer concrete, mixtures AAGC FA.G2 SH2 8M and 10M / AAGC FA.G2 SH3 8M and 10M mixtures have been selected. Results are presented in Table 5.2 and Table 5.3.

Table 5.2.   Influence of NaOH concentration - AAGC FA.G2 SH2 8M and 10M SS2

Mixture	NaOH concentration	Na_2SiO_3/NaOH	Compressive strength (7 days)
AAGC12	8M	1.0	15.8
AAGC15	8M	2.5	20.3
AAGC17	10M	1.0	16.9
AAGC20	10M	2.5	24.3

Table 5.3   Influence of NaOH concentration - AAGC FA.G2 SH3 8M and 10M SS2

Mixture	NaOH concentration	Na_2SiO_3/NaOH	Compressive strength (7 days)
AAGC22	8M	1.0	16.2
AAGC25	8M	2.5	20.5
AAGC26	10M	1.0	17.6
AAGC29	10M	2.5	24.7

Based on the results obtained, it can be stated that the compressive strength of AAGC mixtures increased both with an increase in the concentration of the NaOH solution, but also with an increase in the ratio of alkaline activator solutions. It can be seen that for AAGC mixtures produced using NaOH 8M solution there was an increase in compressive strength of 27%, with the increase in the ratio of the constituent solutions of the alkaline activator, and for those in which NaOH 10M solution was used, the increase was 40%. These results are consistent with the results in the literature and with those obtained for the alkali-activated fly ash-based geopolymer paste presented in the previous chapter.

5.5.2. Influence of NaOH type on the compressive strength of AAGC mixtures

For some of the alkali-activated geopolymer concrete mixtures three types of NaOH has been used to determine its influence on the compressive strength: NaOH flakes, 98%

purity, type SH1 and SH2 and NaOH pearls, 99% purity, type SH3 (Fig. 5.11). The following mixtures have been selected to determine this parameter:

- AAGC FA.G2 SH1 8M SS2;

- AAGC FA.G2 SH2 8M SS2;

- AAGC FA.G2 SH3 8M SS2.

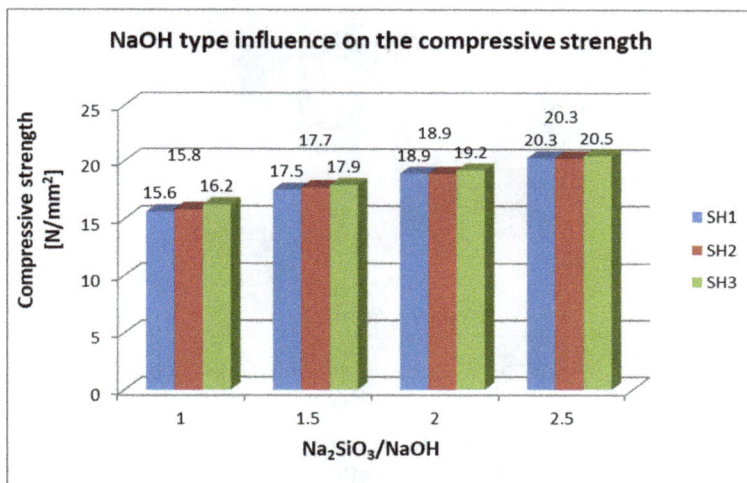

Figure 5.11. Influence of NaOH type on the compressive strength.

Based on the results, it can be seen that the type of sodium hydroxide (NaOH) did not have a major influence on the final compressive strength of the alkali-activated geopolymer concrete mixtures. To the extent that their purity is comparable, the results are similar, there is no major discrepancy, they fall under the measurement uncertainty. One of the reasons users could choose a lower purity of sodium hydroxide might be economic ones, with a lower purity sodium hydroxide being cheaper than a higher purity one.

5.5.3. Influence of Na_2SiO_3/NaOH ratio on the compressive strength of AAGC mixtures

In order to study the influence of the alkaline activators ratio on the compressive strength of the alkali-activated geopolymer concrete (Fig. 5.12 and Fig. 5.13), the following mixtures were analysed: AAGC FA.G2 SH2 8M and 10M SS2, respectively AAGC FA.G2 SH3 8M and 10M SS2.

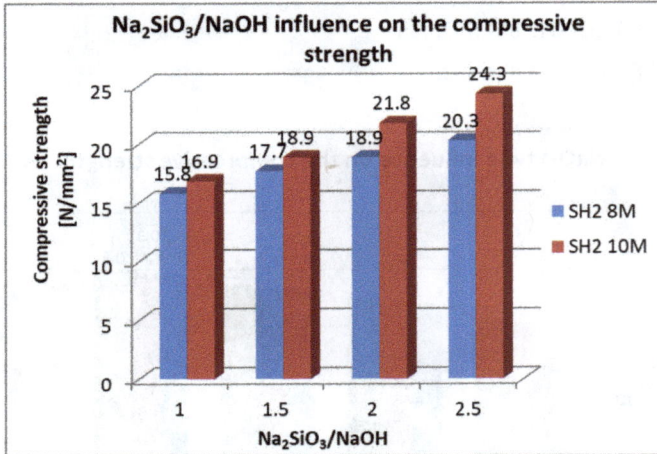

Figure 5.12. $Na_2SiO_3/NaOH$ influence on the compressive strength AAGC FA.G2 SH2 8M and 10M SS2.

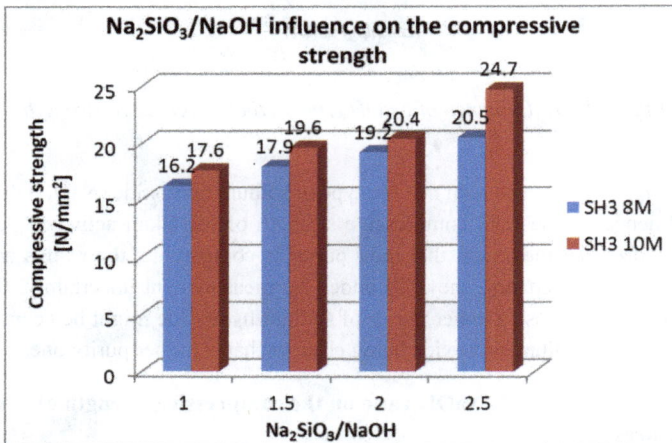

Figure 5.13. $Na_2SiO_3/NaOH$ influence on the compressive strength AAGC FA.G2 SH2 8M and 10M SS2.

Materials Research Forum LLC
https://doi.org/10.21741/9781644901533

The results obtained show that, as in the case of alkali-activated geopolymer paste, as the $Na_2SiO_3/NaOH$ ratio increases the compressive strength of the geopolymer concrete mixtures increased. This is corroborated by the fact that although the workability of the mixtures was lower, their mechanical properties increased.

For all AAGC analyzed mixtures, it can be seen that for a $Na_2SiO_3/NaOH$ ratio of 2.5 the highest compressive strength values are obtained. Also, from Fig. 5.12 and Fig. 5.13 it can be seen that the rate of increase of compressive strength is not constant, this is different for each composition.

Chapter 6

Research Regarding Alkali-Activated Fly Ash-Based Geopolymer Concrete Applications

Although the durability of alkali-activated geopolymer materials is not exactly extensively discussed and analyzed, there are numerous studies regarding the performance of these materials in terms of sustainability [125-127]. Experimental results showed excellent resistance to aging, freeze-thaw, but also to carbonation [128].

The two main barriers to the introduction of new materials in the construction industry [129] are: 1) the need for standards, testing methods, evaluation and admissibility conditions, given that their development and introduction takes a long time and is a gradual process; 2) and the durability performance of such materials, taking into account the fact that they must meet certain requirements, they must withstand for a long time, and for new materials this is not quantified yet.

Based on the results obtained for the alkali-activated fly ash-based geopolymer paste, as well as for geopolymer concrete a number of potential directions for the research on the applicability of the alkali-activated geopolymer material have been observed, two of which have been identified as the most attractive and desirable, both for the study of the parameters of the performance, but also for the delivery of technical know-how in the construction market.

Using the information previously identified as being appropriate and relevant for the production of alkali-activated fly ash-based geopolymer concrete and on the basis of the results obtained, "trial and error" experimental investigations, with the aim of obtaining a construction material made entirely from this material have been carried out. Studies regarding technology transfer to the industry, and the opportunities for optimal adaptation to the existing legislative framework, both at the national and at the European level, for product certification have also been analyzed.

6.1. Alkali-activated fly ash-based geopolymer panels

During the casting of the alkali-activated fly ash-based geopolymer concrete mixtures it was noticed that, when a satisfactory workability is obtained, the material has the capability to take over all forms of a pattern, and after demolding, to obtain a surface that does not require finishing.

Based on preliminary observations made on alkali-activated geopolymer mixtures new applications of the material were found and future aim is to study the possibility of producing prefabricated panel elements, for their use in the established directions (Fig. 6.1 and Fig. 6.2).

Figure 6.1.   AAGC 50 x 25 cm panel.

Figure 6.2.   AAGC 50 x 20 cm panel.

6.2. Alkali-activated fly ash-based geopolymer paving blocks

By using the full information on the production of the alkali-activated geopolymer concrete, as well as on the basis of the obtained results, a "trial and error" experimental research was started, with the aim of producing building geopolymer materials (geopolymer paving blocks), through the study of technology transfer to the industry, by enhancing the legal framework, which sets out the certification of products for the construction market.

To produce alkali-activated geopolymer paving blocks the following materials have been used:

-       Fly ash from Rovinari Power Plant – FA.R;

-       Sodium hydroxide solution – 8M and 10M;

-       Sodium silicate solution – SH2;

-       Natural aggregates, granular class 0/4 mm (S) and 4/8 mm (CA) (sand and coarse aggregates).

The technology of producing alkali-activated geopolymer paving blocks was based on the principles set out for the production of alkali-activated geopolymer concrete and are presented in Fig. 6.3.

Mixing of the materials was made at a constant temperature of $(20\pm2)°C$ and following the next steps:

-       Mixing of the sand and the coarse aggregates (0/4 mm and 4/8 mm) at low speed for 30 seconds;

-       The necessary amount of fly ash was then added and the raw materials (fly ash + aggregates) were mixed together for 30 seconds;

-       The alkaline activator was added during 90 seconds over the homogenous mixture;

-       After adding the total alkaline activator quantity, the mixing was continued at low speed for 3 minutes;

-       Casting into polypropylene molds, with corresponding vibration for 5 minutes and starting of the heat treatment (70°C / 24 hours);

-       After demolding, the geopolymer paving blocks were stored in the climatic chamber at the temperature T $(20\pm1)°C$ and relative humidity RH $(60\pm5)\%$, and the tests for their mechanical properties were determined at 7 days of age.

The molds, casting, maturation and final product are shown in Fig. 6.4 – 6.11.

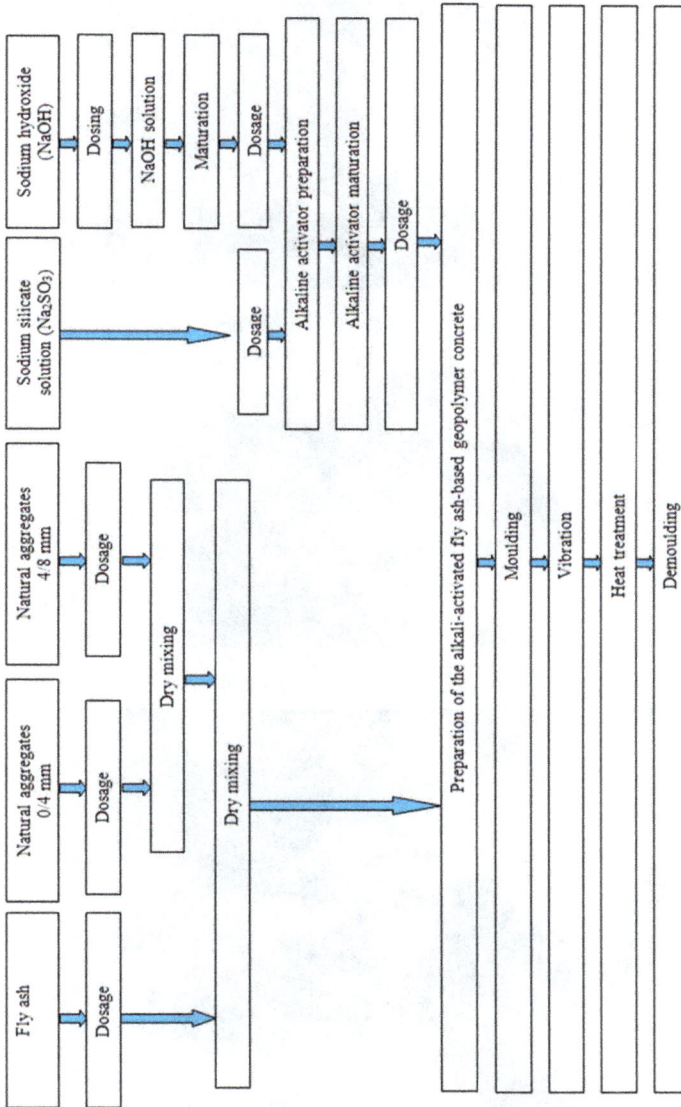

Figure 6.3.   Technological flow for the production of geopolymer paving blocks.

Figure 6.4.   23.7 x 10.3 x 6.0 cm mould.

Figure 6.5.   22.5 x 8.8 x 6.0 cm mould.

Figure 6.6.   Moulding of the paving blocks.

Figure 6.7.   23.7 x 10.3 x 6.0 cm paving blocks after vibration.

Figure 6.8.   22.5 x 8.8 x 6.0 cm paving blocks after vibration.

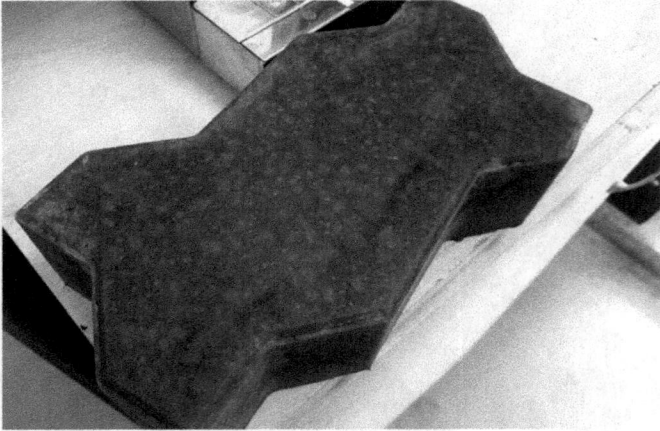

Figure 6.9.   23.7 x 10.3 x 6.0 cm paving blocks.

Figure 6.10. 22.5 x 8.8 x 6.0 cm paving blocks.

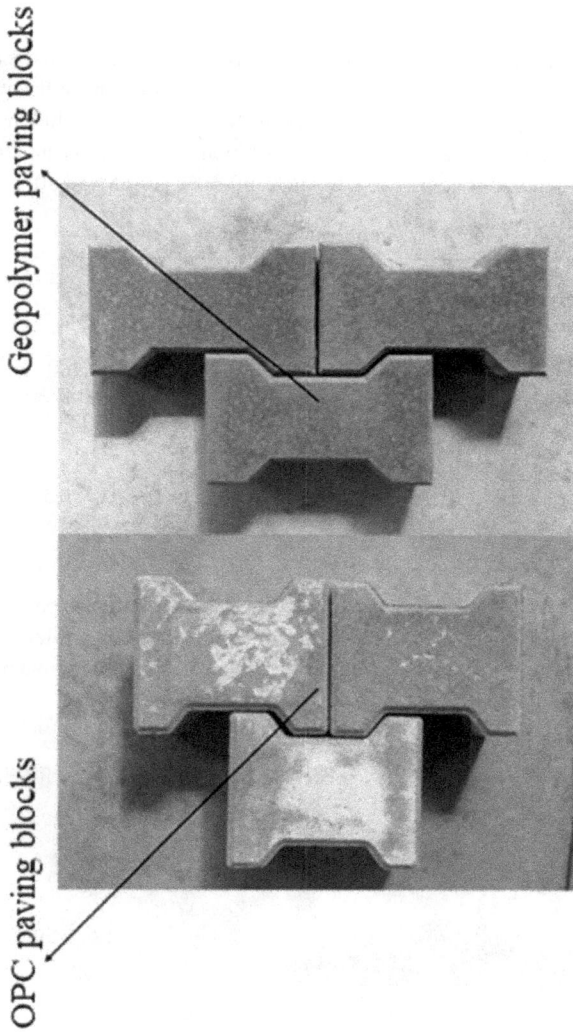

Figure 6.11. OPC paving blocks vs. Geopolymer paving blocks.

Materials Research Forum LLC
https://doi.org/10.21741/9781644901533

Mix-design parameters

In order to study the parameters that affect the mechanical properties of the alkali-activated geopolymer paving blocks, the following preliminary mixtures have been produced, with the following constant parameters: AA/FA ratio, Na_2SiO_3/NaOH solution ratio and S/CA ratio. The variable parameters were the NaOH solution concentration and the ratio between fly ash and total amount of aggregates (50-50%, 40-60%, 35-65% şi 30-70%) (Table 6.1).

Table 6.1.   Geopolymer paving blocks mix-design

Material / Parameter	I	II	III	IV	V
Fly ash	1.0	1.0	1.0	1.0	1.0
AA/FA	0.9	0.9	0.9	0.9	0.9
Na_2SiO_3	0.64	0.64	0.64	0.64	0.64
NaOH	0.26	0.26	0.26	0.26	0.26
NaOH conc.	8M	8M	8M	8M	10M
Na_2SiO_3 / NaOH	2.5	2,5	2,5	2,5	2.5
S 0/4 mm	0.5	0,75	0,93	1,17	0.93
CA 4/8 mm	0.5	0,75	0,93	1,17	0.93
S : CA	0.5 : 0.5	0.5 : 0.5	0.5 : 0.5	0.5 : 0.5	0.5 : 0.5
Binder / Aggregate	1 : 1	1 : 1.50	1: 1.85	1 : 2.33	1: 1.85

To establish the optimal mix-design parameters of the geopolymer paving blocks and to determine their mechanical and durability performances, at 2 days (after finishing the heat treatment), tensile splitting strength was performed on all samples to obtain information on the geopolymer concrete matrix (Fig. 6.12).

Figure 6.12. Geopolymer paving blocks sections.

It can be seen from Fig. 6.12 that the arrangement of aggregates in the alkali-activated geopolymer concrete matrix is strongly influenced by the total amount of geopolymer binder. At an equal amount of binder – aggregates, a separation of the 4/8 mm aggregates can be observed. As the amount of aggregates in the mixture increased, a better dispersion can be observed, hence a more homogeneous matrix. Preliminary results obtained on the geopolymer paving blocks when performing the tensile splitting strength tests shown that mixtures with the ratio binder-aggregates 35%-65% had the highest values, therefore it has been chosen the optimum mix-design in order to establish further mechanical and durability properties.

Based on the results obtained regarding the possibility of developing such construction elements, research was started on the physical, mechanical and durability characteristics of the products obtained, in order to be able to study their potential for different areas of applicability:

- Weathering resistance – total water absorption;

- Weathering resistance – freeze-thaw resistance with de-icing salt;

- Tensile splitting strength;

- Abrasion resistance;

- Slip/skid resistance.

All the tests were performed according to EN 1338 *Concrete paving blocks.* Requirements and test methods [130] which specifies the materials, characteristics, conditions and methods of testing paving blocks, and is applicable for the use of pedestrians, the use of vehicles, bicycle lanes, parking lots, roads, highways, industrial areas, etc. To be used in specific applications, they must comply with certain conditions at the time of their declaration as fit for use by the manufacturer.

The results of the experimental tests and the admissibility conditions according to the intended area of use obtained on the alkali-activated fly ash-based geopolymer paving blocks are presented in Table 6.2. Instances from the experimental tests are shown in Fig. 6.13-6.19.

Table 6.2.   Experimental test results and admissibility conditions

Characteristic Testing standard [Measurement unit]	Average results	Standard EN 1338 Admissibility conditions
Apparent density, 7 days EN 12390-7 [Kg/m^3]	1870	-
Water absorbtion Annex E - EN 1338 [%]	5.6	Class 1, Marking A, No measured performance
		Class 2, Marking B, ≤ 6 % as average value
Freeze-thaw resistance with de-icing salt Annex D - EN 1338 Surface mass lost per unit area [kg/m^3]	0.6	Class 3, Marking D, ≤ 1.0 as average value, with no individual value > 1.5
Böhme abrasion resistance EN 1338 Loss in volume [mm^3 / 5000 mm^2]	16993	Class 1, Marking F, No measured performance
		Class 3, Marking H, $\leq 20\ 000$ mm^3 / 5000 mm^2
		Class 4, Marking I, $\leq 18\ 000$ mm^3 / 5000 mm^2
Unpolished slip resistance (USVR) Annex I - EN 1338	38	Moderate slip potential readings on the slip resistance Test falling within the range of 20-39, Framing Class X
Tensile splitting strength EN 1338 T [Mpa] F [N/mm]	Mixture V 3.8 410	Pt. 5.3.3 EN 1338 T ≥ 3.6 Mpa F ≥ 250 N/mm

Alkali-activated geopolymer paving blocks have been subjected to tests for mechanical and durability performance by assimilating standard EN 1338. The results confirmed their ability to comply with this standard, for each of the requirements imposed by Annex ZA [130] of the standard, which demonstrates the possibility of their use for paving elements in the construction materials industry. Moreover, by optimizing the mixtures, improved performance can be achieved, attesting their use for the destinations specified by this standard.

Figure 6.13. Sample used for the abrasion resistance test.

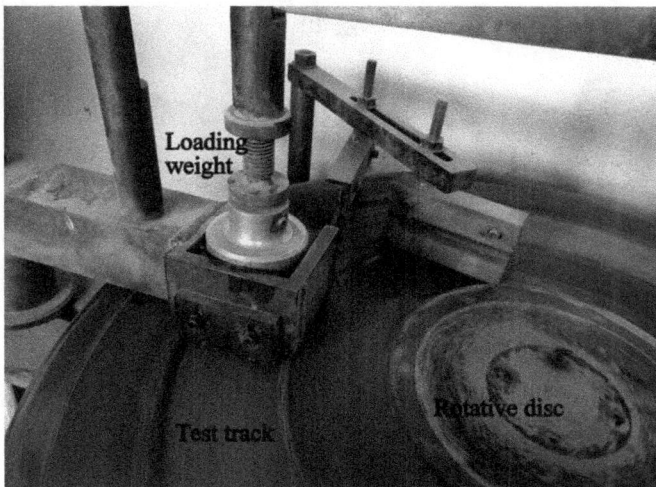

Figure 6.14. Placing of the samples on the abrasion disc.

Figure 6.15. Paving block sample before and after Bohme abrasion test.

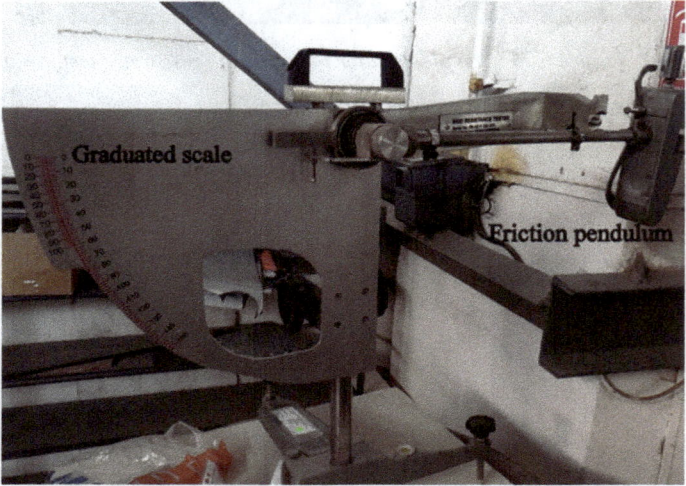

Figure 6.16. Slip resistance measuring equipment.

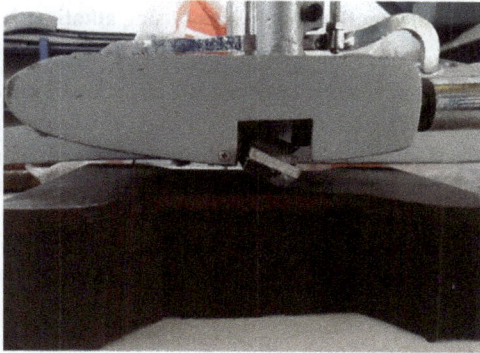

Figure 6.17. Unpolished Slip Resistance (USVR).

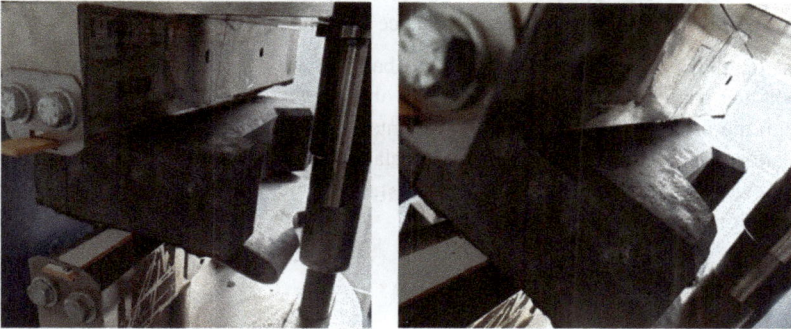

Figure 6.18. Placement of the geopolymer paving blocks for tensile splitting test.

Figure 6.19. Geopolymer paving block after the tensile splitting test.

6.3. Legislative framework on the possibility of using alkali-activated geopolymer paving blocks

As the results of the research are transferred to the industry, in order to be able to successfully use a building material, two ways of transfer are distinguished. The attestation of conformity of products intended to be used in the construction industry is a procedural system by which a notified body assesses and determines the compliance of products with technical specifications, applicable for use in the industrial sector [131]. Harmonized technical specifications are documents comprising European or national standards as well as European assessment documents. Construction products for which a harmonized technical specification exists are covered by EU Regulation no. 305/2011 for the certification of those materials [132]. For products not covered by a harmonized technical specification, the manner in which they are placed on the market shall be determined by a Technical Approval - defined as the favorable technical assessment, which attests the ability to use a building material in accordance with the legal requirements.

Based on the results obtained and on the basis of future research on alkali-activated geopolymer paving blocks, the possibility of obtaining a technical approval is considered, through the production of technical documentation, which shall contain the information, evaluations, experimental research, etc., in relation to the technical specifications relating to the evaluation of the conformity of the geopolymer paving blocks, for use in the construction industry.

Materials Research Forum LLC
https://doi.org/10.21741/9781644901533

Conclusions and Final Remarks

The obtained and presented results offer an analysis of the current state of research in the field of alkali-activated fly ash-based geopolymer materials, by presenting the fundamental aspects underlying the production of this type of material, the characteristics that the constituent materials must meet, the factors of influence and the possibilities of extending laboratory research to industry, by presenting the most relevant directions of applicability of the material.

The fundamentals of geopolymers and geopolymerization were compiled together with several important factors, affecting their properties and characteristics: the chemical composition of the raw material, the alkaline activator, the heat treatment method, etc.

Most of the alkali-activated materials' properties were analyzed by comparing them with the performance of ordinary Portland cement concrete, even if the production of this type of materials (namely, geopolymers), binders or mortars and concretes, is totally different from the classic and usual ones. The main properties of the geopolymer materials obtained by appropriate manufacturing processes and heat treatment are very much dependent on the characteristics of the materials used. The multitude of variables that intervene in the production of such materials make alkali-activated geopolymer materials the hypothesis of thorough studies, in which all these variables must be taken into account.

Overall, trends in the field of building materials, civil engineering and architecture should focus on understanding the mechanisms of geopolymerization and the need to develop regulations for the production of these types of materials and to be able to develop this material for commercial use. In order for this to be achievable, in-depth studies in the field and the finding of common elements are necessary, in order to transform this material from a niche application into a generalized one.

An important part of the research was dedicated to the identification, both of the way of realization of the alkali-activated fly ash-based geopolymer binder, and the presentation of the specific dosages used for its production, which leads to a better understanding of the mechanism of production of the geopolymerization reaction. 47 alkali-activated fly ash-based geopolymer binder mixture, grouped into 11 categories of influencing factors, developed and evaluated unitary, generated valuable experimental results that substantiated relevant conclusions on the principles of development of these materials in common practice.

The sodium hydroxide solution, an essential component of the alkaline liquid, must be properly prepared at least 24 hours before the alkaline mixture is made. The component materials require precise weighing, and after making the solution, it must be stored in hermetically sealed containers, in order to avoid loss of its chemical properties, since there is a risk of carbonation when in contact with oxygen in the room.

The alkaline liquid is prepared at least 24 hours before the mixing starts, with strict compliance with the proposed ratios for specific mixtures. Its storage will also be carried out in hermetically sealed containers, and the combination of the two components (sodium silicate solution and sodium hydroxide solution) will always be carried out by forced mixing, in order to boost their reaction.

The casting of the alkali-activated fly ash-based geopolymer materials requires careful preparation of the molds, by thoroughly lubricating them with form release agent solution, since it has been found that the geopolymer paste has a strong adhesion with the metal.

The heat treatment to which the geopolymer paste samples are subjected requires the mandatory protection of the samples by glass coating, to avoid a rapid evaporation of water from the mixtures, against the background of high temperatures, which leads to the degradation of mechanical resistances.

The mechanisms that produce the geopolymerization reaction and the oxides that enter its composition and that lead to the formation of the geopolymer paste, require high knowledge and understanding, they represent very important elements in the production of binders with high mechanical properties. These mechanisms are essentially related to the molar ratios of the constituents of each of the materials that have entered the composition of the geopolymer; based on these considerations, each mixture also assumed the recording of these molar ratios, to substantiate consistent and significant comparative analyses in this regard.

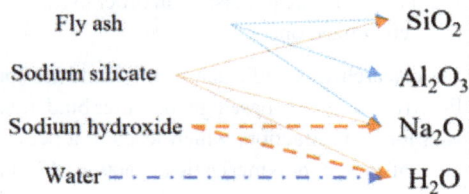

Origin of the molar elements that produce the alkali-activated geopolymer binder.

The SiO_2 / Al_2O_3 ratio has an important influence on the mechanical strength of alkali-activated geopolymer materials. This factor is synthetically quantified as the additional amount of Silicon (Si) required for the mixtures, generated by the alkaline activator, and determined by knowing the chemical composition of the source material (in this case, fly ash) as the main Si species supplier.

The H_2O / Na_2O ratio is an important indicator of the total amount of water in the mixture, at the time the other parameters are known, and corresponds to the alkaline concentration in the activator.

The ratio of liquids to solids in the mixture (w/s) also represents an important factor in the successful production of alkali-activated geopolymer binders.

The results obtained on the alkali-activated fly ash-based geopolymer binders were analyzed comparatively and generated significant conclusions on the different influences that both the constituent materials can have on the final mechanical properties of the material, but also the specific ratios in which they are used in mixtures. Consequently, the determining parameters in the development high-performance geopolymer mixtures can be identified:

The chemical and physical properties of fly ash are defining for the development of alkali-activated geopolymer mixtures. Thus, its oxidic composition is essential for the realization of the geopolymer binder, together with its *fineness* (characterized by the curve of granularity or rest on the sieve of 0.045 mm), with substantial influences on the characteristics of the final product, both in fresh (workability) but also in the hardened state (mechanical performance and durability). As the fineness of the thermal power plant ash increases, the compressive strength of the geopolymer mixtures also increase, observing a decrease in the need for alkaline activator for the development of the geopolymerization reaction.

The properties of the alkaline activator and the specific dosages have a consistent influence on the compressive strength of the alkali-activated geopolymer binder, which generally increases with the increase in the specific concentration of alkaline activators. A higher concentration of the NaOH solution can give rise to stronger Si-O-Si bonds and generally improves the dissolution of the source materials in the presence of activators, which is observed by increasing the mechanical strength of the material. The experimental analysis also shows that as the ratio of the constituent solutions of the alkaline activator (Na_2SiO_3/NaOH) increases, the mechanical properties of the alkaline activated geopolymer paste also improve.

The hardened alkali-activated geopolymer binder, as a result of the heat treatment stage, represents a homogeneous material, gray-brown-dark, characterized by the development of the following physical-mechanical performance (values recorded at the age of 7 days after preparation) that vary in the following intervals:

Materials Research Forum LLC
https://doi.org/10.21741/9781644901533

- Apparent density: 1200 - 1400 kg/m^3;

- Flexural strength: 1.3 - 6.5 N/mm^2;

- Compressive strength: 5.1 - 35.0 N/mm^2.

The present scientific approach clearly emphasizes that the multitude of factors that influence the production of alkali-activated geopolymer binder must be taken into account in the design of this type of material, and subsequently by specific optimization of the mixtures, depending on the performance requirements (final strength and durability); thus, profile applications are identified, through the transition to alkaline-activated geopolymer mortars and even alkaline-activated geopolymer concretes.

The results obtained for the production of alkali-activated fly ash-based geopolymer concrete, as well as the results obtained on the influences affecting this type of material, in accordance with the specialized literature, confirm the results obtained on alkali- activated geopolymer paste and generate a complex and current view on the actual potential for production and use of this material in Romania (production technology and application fields).

Through the optimization of the mixtures the viability of the material and the production of the alkali-activated fly ash-based geopolymer binder was obtained (geopolymer paving blocks), measured from the point of view of their mechanical properties and durability. This opens up new opportunities for the development of this type of material in the future, through an ongoing process of compositional optimization, by adapting to the characteristics of the actual characteristics of the source material, and by the start of the procedures required to make use of the material in the building materials market.

Using all the information obtained about the production of alkali-activated fly ash-based geopolymer binders, the present experimental "trial-and-error " program was started, with the aim of obtaining and characterizing this innovative product, using materials available on a national level.

Geopolymer technology is gaining consistent ground due to the successful implementation, in certain countries, of this type of concrete, in different areas. This aspect is mainly due to the need to align with the principles of the circular economy through the recovery of ash and the prevention of waste generation.

The need to continue existing studies is identified due to the many unknowns in the field regarding the geopolymerization process and the properties of geopolymer concrete, as well as products derived from this process. Thus, this research direction fits into the current complex theme, aligned with worldwide research on innovative ways of harnessing fly ash as a raw material in the development of new building materials.

The experimental results provide the basis for the validation of the possibility of using fly ash, from local sources in Romania, for the development of alkali-activated geopolymer materials (paste, mortar and geopolymer concrete).

Moreover, the possibility of integrating geopolymer, paste and concrete mixtures into building materials and products has been demonstrated; this fact, however, is only possible through compositional optimization processes, the multitude of influences that affect this type of material.

The research undertaken generates valuable conclusions for the development of fly ash-based geopolymer materials and their effective integration in the actual current interest of the industry profile:

- the physical and chemical characteristics of the constituent materials in the geopolymer material mixtures are of essential importance for the production of this type of material;

- the instability of the chemical composition of the fly ash and the differences in physical characteristics also have a huge influence on the process of geopolymerization;

- the production technology of the alkali-activated fly ash-based geopolymer materials and the optimal dosages require careful selection of the materials in order to generate a complete and efficient geopolymerization process, as well as to obtain the expected mechanical properties;

- increasing the concentration of sodium hydroxide solution (in molar terms) causes higher values of compressive strength;

- increasing the ratio of sodium silicate solution to sodium hydroxide solution increases the compressive strength performance of the alkali-activated geopolymer material;

- temperature and duration of heat treatment are important factors in the development of the mechanical strengths of geopolymer materials. It is concluded that a heat treatment at 70°C for 24 hours is optimal for achieving the projected performance of the geopolymer material;

- due to the use of heat treatment for the production of geopolymer materials, it is found that the values of compressive strength stabilize at the age of 7 days, the subsequent increases up to the age of 28 days are negligible;

- quite high mechanical strength, achieved at a young age (7 days), proves the applicability of this material in the manufacture of prefabricated construction products;

- the raw material, especially fly ash available in Romania can be successfully used in the production of alkali-activated fly ash-based geopolymer binders.

By addressing a topic of current research, from the point-of-view of the concepts of "sustainable development" and "circular economy", through the re-introduction into the economic cycle of the fly ash, which is in a sub-product of industry and becoming the raw material in the production of alkali-activated materials, thus preventing the conversion to stored waste and by the selection and adoption of specific laboratory procedures for the production of alkali-activated fly ash-based geopolymer materials, the possibility of using this sub-product for the production of geopolymer materials was demonstrated.

Based on the results obtained, the authors appreciate that all the established research objectives regarding the study and production of these materials have been met. The production of alkali-activated fly ash-based geopolymer materials is an area that requires great attention, both due to the specific procedures, but also due to the increasing interest for ecological materials.

References

[1] V.M. Malhotra, Making Concrete "Greener" with Fly Ash, ACI Concr. Int. 21(5) (1999) 61-66.

[2] IEA, 2009. Cement Technology Roadmap: Carbon Emissions Reductions up to 2050, IEA Technology Roadmaps. Paris: OECD Publishing.

[3] Information on http://cembureau.eu/media/ms1ilpgv/19901-cembureau-position-paper-on-the-public-consultation-on-a-new-eu-forest-strategy.pdf

[4] H. Szilagyi, Betoane speciale, Napoca Star, Cluj-Napoca, Romania, 2011.

[5] J. Davidovits, 1999. Chemistry of Geopolymeric Systems. Terminology. Saint-Quentin: Geopolymer Institute.

[6] A. Palomo, M.W. Grutzeck, M.T. Blanco, Alkali-Activated Fly Ashes. A Cement for the Future, Cem. Conc. Res. 29(8) (1999) 1323-1329. https://doi.org/10.1016/S0008-8846(98)00243-9

[7] N.A. Lloyd, B.V. Rangan, Geopolymer concrete with fly ash, Second International Conference on Sustainable Materials and Technologies, Italy, 2010.

[8] Information on https://www.vgb.org/vgbmultimedia/PT202010HARRIS-p-16422.pdf

[9] BP Statistical Review of World Energy, 67th edition, June 2018.

[10] A.M. Mustafa Al Bakri, H. Kamarudin, M. Bnhussain, I. Khairul Nizar, W.I.W. Mastura, Mechanism and Chemical Reaction of Fly Ash Geopolymer Cement - A Review, J. Asian Sci. Res. 1(5) (2011) 247-253.

[11] A.M. Mustafa Al Bakri., K. Hussin, M. Bnuhussain, K.N. Ismail, M.I. Ahmad, Chemical Reactions in the Geopolymerisation Process Using Fly Ash-Based Geopolymer: A review, AJBAS 5(7) (2011) 1199-1203.

[12] P. De Silva, K. Sagoe-Crenstil, V. Sirivivatnanon, Kinetics of geopolymerisation: Role of Al2O3 and SiO2, Cem. Conc. Res. 37(4) (2007) 512-518. https://doi.org/10.1016/j.cemconres.2007.01.003

[13] P. Duxon, A. Fernandez-Jimenez, J.L. Provis, G.C. Lukey, A. Palomo, J.S.J. van Deventer, Geopolymer technology: the current state of the art, J.Mater. Sci. 42(9) (2007) 2917-2933. https://doi.org/10.1007/s10853-006-0637-z

[14] J. Davidovits, Geopolymers of the first generation: SILIFACE-Process, Geopolymer '88, First Europrean Conference on Soft Mineralogy, Compiegne, France (1988) 49-67.

[15] F. Pacheco-Torgal, J. Castro-Gomez, S. Jalali, Alkali-activated binders: A review Part 1. Histrorical background terminology, reaction mechanisms and hydration products, J. Constr. Build. Mater. 22 (2008) 1305-1314. https://doi.org/10.1016/j.conbuildmat.2007.10.015

[16] F. Skvara, Alkali activated material- geopolymer, Department of Glass and Ceramics, Faculty of Chemical Technology, ICT Prague (2007) 661-676.

[17] A. Buchwald, What are geopolymers? Current State of Research and Technology, The Opportunities They Offer, and Their Significance for The Precast Industry, Concrete Precasting Plant and Technology 72(7) (2008) 42-49.

[18] B. Mehdi, Geopolymer thechnology, from fundamentals to advanced applications: a review, Mater. Technol. 24(2) (2009) 79-87. https://doi.org/10.1179/175355509X449355

[19] J.G.S. Van Jaarveld, J.S.J. van Deventer, G.C. Lukey, The characterisation of source materials in fly ash-based geopolymers, Mater. Lett. 57(7) (2003) 1272-1280. https://doi.org/10.1016/S0167-577X(02)00971-0

[20] J. Davidovits, Geopolymeric inorganic polymeric new materials, J. Mater. Education 16 (1994) 91-139.

[21] H. Wang, H. Li, F. Yan, Synthesis and mechanical properties of metakaolinite-based geopolymer, Colloids Surf. A: Phys. Eng. Asp. 268 (2005) 1-6. https://doi.org/10.1016/j.colsurfa.2005.01.016

[22] Komnitsas K., Zaharaki D., Perdiatsis V. (2009), Efect of synthesis parameters on the compressive strength of low-calcium ferronickel slag inorganic polymers, J. Haz. Mater. 161:760-768. https://doi.org/10.1016/j.jhazmat.2008.04.055

[23] P. Duxon, J.L. Provis, G.C. Lukey, S.W. Maillicoat, V.M. Kriven, J.S.J. van Deventer, Understanding the relationship between geopolymer composition, microstructure and mechanical properties, Colloids Surf. A: Physicochem. Eng. Aspects 269(1-3) (2005) 47-58. https://doi.org/10.1016/j.colsurfa.2005.06.060

[24] A. Fernández-Jiménez, A. Palomo, Composition and microstructure of alkali activated fly ash binder: effect of the activator, Cem. Conc. Res. 35(10) (2005) 1984-1992. https://doi.org/10.1016/j.cemconres.2005.03.003

[25] C. Panagiotopoulou, E. Kontori, T. Perraki, G. Kakali, Dissolution of aluminosilicate minerals and by-products in alkaline media, J. Mater. Sci. 42(9) (2007) 2967-2973. https://doi.org/10.1007/s10853-006-0531-8

[26] D. Hardjito, C. Cheak, C. Ing, Strength and setting times of low calcium fly ash-

Materials Research Forum LLC
https://doi.org/10.21741/9781644901533

based geopolymer mortar, Mod. Appl. Sci. 2(4) (2008) 3-11.
https://doi.org/10.5539/mas.v2n4p3

[27] D. Dimas, I. Giannopoulou, D. Panias, Polymerization in sodium silicate solutions: a fundamental process in geopolymerization technology, J. Mater. Sci. 44(14) (2009) 3719-3730. https://doi.org/10.1007/s10853-009-3497-5

[28] K. Vijai, R. Kumutha, B.G. Vishnuram, Effect of types of curing on strength of geopolymer concrete, I. J. Phys. Sci. 5(9) (2010) 1419-1423.

[29] P. Mehta, Greening of the Concrete Industry for Sustainable Development, ACI Conc. Int. 24(7) (2002) 23-28.

[30] A. Mishra, D. Choudhary, N. Jain, M. Kumar, N. Sharda, D. Dutta, Effect of concentration of alkaline liquid and curing time on strength and water absorption of geopolymer concrete, ARPN J. Eng. Appl. Sci. 3(1) (2008) 14-18.

[31] V.F.F. Barbosa, K.J.D. MacKenzie, C. Thaumaturgo, Synthesis and Characterisation of Materials Based on Inorganic Polymers of Alumina and Silica: Sodium Polysialate Polymers, Int. J. Inorganic Mater. 2(4) (2000) 309-317. https://doi.org/10.1016/S1466-6049(00)00041-6

[32] D.M. Roy, Alkali-Activated Cements, Opportunities and Challenges. Cem. Conc. Res. 29(2) (1999) 249-254 https://doi.org/10.1016/S0008-8846(98)00093-3

[33] Z. Xie, Y. Xi, Hardening mechanism of an alkali activated class F fly ash, Cem. Conc. Res. 31(9) (2001) 1245-1249. https://doi.org/10.1016/S0008-8846(01)00571-3

[34] D. Panias, I.P. Giannopoulou, T. Perraki, Effect of synthesis parameters on the mechanical properties of fly ash-based geopolymers, Colloids Surf. A Physicochem. Eng. Asp. 301(1-3) (2007) 246-254. https://doi.org/10.1016/j.colsurfa.2006.12.064

[35] P. Sukmak, S. Horpibulsuk, S.-L. Shen, P. Chindaprasirt, C. Suksiripattanapong, Factors influencing strength development in clay-fly ash geopolymer, Constr. Build. Mater. 47 (2013) 1125-1136. https://doi.org/10.1016/j.conbuildmat.2013.05.104

[36] A. Lăzărescu, H. Szilagyi, A. Ioani, C. Baeră, Parameters Affecting the Mechanical Properties of Fly Ash-Based Geopolymer Binders - Experimental Results, IOP Conf. Ser: Mater. Sci. Eng. 374 (2018) https://doi.org/10.1088/1757-899X/374/1/012035

[37] A. Lăzărescu, H. Szilagyi, A. Ioani, C. Baeră, Effect of sodium hydroxide

concentration and alkaline activator ratio on the mechanical properties of fly ash-based geopolymer binders, Proceedings of the 12th International PhD Symposium in Civil Engineering, Czech Technical University in Prague, (2018) 101-107.

[38] A. Lăzărescu, C. Mircea, H. Szilagyi, C. Baeră, Mechanical properties of alkali activated geopolymer paste using different Romanian fly ash sources-experimental results, MATEC Web Conf. 289 (2019) 11001.
https://doi.org/10.1051/matecconf/201928911001

[39] A. Lăzărescu, H. Szialgyi, C. Baeră, A. Ioani, The Effect of Alkaline Activator Ratio on the Compressive Strength of Fly Ash-Based Geopolymer Paste, IOP Conf. Ser: Mater. Sci. Eng. 209 (2017). https://doi.org/10.1088/1757-899X/209/1/012064

[40] P. Nath, P.K. Sarker, W.K. Biswas, Effect of fly ash on the service life, carbon footprint and embodied energy of high strength concrete in the marine environment, Energy Build. 158 (2018) 1694-1702.
https://doi.org/10.1016/j.enbuild.2017.12.011

[41] S. Taebuanhuad, U. Rattanasak, S. Jenjirapanya, Strength behavior of fly ash geopolymer with microwave pre-radiation curing, J. Ind. Technol. 8(2) (2012) 1-8.

[42] D. Dutta, S. Chakrabarty, C. Bose, S. Ghosh, Evaluation of geopolymer properties with temperature imposed on activator prior mixing with fly ash, Int. J. Civ. Struct. Eng. 3(1) (2012) 205-213.

[43] J.G.S. van Jaarveld, J.S.J. van Deventer, G.C. Lukey, The Effect of Composition and Temperature on the Properties of Fly Ash and Kaolinite-based Geopolymers, Chem. Eng. J. 89(1-3) (2002) 63-73. https://doi.org/10.1016/S1385-8947(02)00025-6

[44] P. Rovnaník, Effect of curing temperature on the development of hard structure of metakaolin-based geopolymer, Constr. Build. Mater. 24(7) (2010) 1176-1183.
https://doi.org/10.1016/j.conbuildmat.2009.12.023

[45] S. Demie, M. Nuruddin, M. Ahmed, N. Shafiq, Effects of curing temperature and superplasticizer on workability and compressive strength of self-compacting geopolymer concrete, National Postgraduate Conference (NPC), Tronoh Perak, Malaysia, IEEE (2011) 1-5. https://doi.org/10.1109/NatPC.2011.6136362

[46] D. Reddy, J. Edouard, K. Sobhan, Durability of fly ash-based geopolymer structural concrete in the marine environment, J. Mater. Civ. Eng. 25(6) (2012) 781-787.
https://doi.org/10.1061/(ASCE)MT.1943-5533.0000632

[47] K. Komnitsas, D. Zaharaki, V. Perdikatsis, Geopolymerisation of low calcium

ferronickel slags, J. Mater. Sci. 42(9) (2004) 3073-3082.
https://doi.org/10.1007/s10853-006-0529-2

[48] B. Chatveera, N. Makul, Properties of geopolymer mortar produced from fly ash and rice husk ash: Influences of fly ash-rice husk ash ratio and Na2SiO3-NaOH ratio under curing by microwave energy, KMUTT Research & Development Journal 35(3) (2012) 299-309.

[49] P. Chindaprasirt, T. Chareerat, V. Sirivivatnanon, Workability and strength of coarse high calcium fly ash geopolymer, Cem. Con. Comp. 29(3) (2007) 224-229.
https://doi.org/10.1016/j.cemconcomp.2006.11.002

[50] Z. Zhang, X. Yao, H. Zhu, Y. Chen, Role of water in the synthesis of calcined kaolin-based geopolymer, App. Clay Sci. 43(2) (2009) 218-223.
https://doi.org/10.1016/j.clay.2008.09.003

[51] T. Phoo-ngernkham, T. Sinsiri, Workability and compressive strength of geopolymer mortar from fly ash containing diatomite, KKU Eng. J. 38(1) (2011) 11-26.

[52] T. Bakharev, Geopolymeric materials prepared using class F fly ash and elevated temperature curing, Cem. Conc. Res. 35(6) (2005) 1224-1232.
https://doi.org/10.1016/j.cemconres.2004.06.031

[53] K. Komnitsas, D. Zaharaki, Geopolymerisation: A review and prospects for the minerals industry, Minerals Eng. 20(14) (2007) 1261-1277.
https://doi.org/10.1016/j.mineng.2007.07.011

[54] A. Fernández-Jiménez, I. Garcia-Lodeiro, A. Palomo, Durability of alkali-activated fly ash cementitious materials, J. Mater. Sci. 42(9) (2007) 3055-3065.
https://doi.org/10.1007/s10853-006-0584-8

[55] M. Sofi, J.S.J. van Deventer, P. Mendis, G.C. Lukey, Engineering properties of inorganic polymer concretes (IPCs), Cem. Conc. Res. 37(2) (2007) 251-257.
https://doi.org/10.1016/j.cemconres.2006.10.008

[56] S. Wallah, Drying shrinkage of heat-cured fly ash-based geopolymer concrete, Modern App. Sci. 3(12) (2009) 14-21. https://doi.org/10.5539/mas.v3n12p14

[57] J. Davidovits, 30 years of successes and failures in geopolymer applications: Market trends and potential breakthroughs, Geopolymer 2002 3rd International Conference, Melbourne Australia, 2020.

[58] M. Lizcano, A. Gonzalez, S. Basu, K. Lozano, M. Radovic, Effects of water content and chemical composition on structural properties of alkaline activated metakaolin-based geopolymers, J. Am. Ceram. Soc. 95(7) (2012) 2169-2177.

https://doi.org/10.1111/j.1551-2916.2012.05184.x

[59] Winston R. (2008), Corrosion and Corrosion Control, 4th ed. West Sussex: Wiley, UK.

[60] A. Aleem, P.D. Arumairaj, Optimum mix for the geopolymer concrete, Indian J. Sci. Tech. 5(3) (2012) 2299-2301. https://doi.org/10.17485/ijst/2012/v5i3.8

[61] P. Chindaprasirt, T. Chareerat, S. Hatanaka, T. Cao, High-strength geopolymer using fine high-calcium fly ash, J. Mater. Civ. Eng. 23(3) (2010) 264-270. https://doi.org/10.1061/(ASCE)MT.1943-5533.0000161

[62] J. Petermann, A. Saeed, M. Hammons, Alkali-Activated Geopolymers: A Literature Review, Panama City, USA: Applied Research Associates, Inc. 2010. https://doi.org/10.21236/ADA559113

[63] S. Kumar, R. Kumar, Mechanical activation of fly ash: Effect on reaction, structure and properties of resulting geopolymer, Ceram. Int. 37(2) (2011) 533-541. https://doi.org/10.1016/j.ceramint.2010.09.038

[64] K. Somna, C. Jaturapitakkul, P. Kajitvichyanukul, P. Chindaprasirt, NaOH-activated ground fly ash geopolymer cured at ambient temperature, Fuel 90(6) (2011) 2118-2124. https://doi.org/10.1016/j.fuel.2011.01.018

[65] J. Temuujin, A. van Riessen, Effect of fly ash preliminary calcination on the properties of geopolymer, J. Hazard. Mater. 164(2) (2009) 634-639. https://doi.org/10.1016/j.jhazmat.2008.08.065

[66] T. Phoo-ngernkham, P. Chindaprasirt, V. Sata, S. Hanjitsuwan, S. Hatanaka, The effect of adding nano-SiO2 and nano-Al2O3 on properties of high calcium fly ash geopolymer cured at ambient temperature, Mater. Des. 55 (2014) 58-65. https://doi.org/10.1016/j.matdes.2013.09.049

[67] Gourley J.T. (2003), Geopolymers; Opportunities for Environmentally Friendly Costruction Materials, Materials 2003 Conference: Adaptive Materials for a Modern Society, Sydney, Institute of Materials Engineering, Australia.

[68] S. Alonso, A. Palomo, Alkaline activation of metakaolin and calcium hydroxide mixtures: influence of temperature, activator concentration and solids ratio. Mater. Lett. 47(1) (2001) 55-62. https://doi.org/10.1016/S0167-577X(00)00212-3

[69] A. Palomo, M.W. Grutzeck, M.T. Blanco, Alkali-Activated Fly Ashes. A Cement for the Future, Cem. Conc. Res. 29(8) (1999) 1323-1329. https://doi.org/10.1016/S0008-8846(98)00243-9

[70] D. Hardjito, B.V. Rangan, 2005. Development and properties of low-calcium fly ash-based geopolymer concrete. Perth, Australia: Curtin Technical University. Retrieved from: https://www.researchgate.net/publication/228794879_Development_and_Propertie s_of_Low-calcium_Fly_Ash_Based_Geopolymer_Concrete

[71] J.G.S. van Jaarveld, J.S.J. van Deventer, G.C. Lukey, The Effect of Composition and Temperature on the Properties of Fly Ash and Kaolinite-based Geopolymers, Chem. Eng. J. 89(1-3) (2002) 63-73. https://doi.org/10.1016/S1385-8947(02)00025-6

[72] A.S. de Vargas, D.C.C. dal Molin, A.C.F. Vilela, F.D. Silva, B. Pavao, H. Veit, The effects of Na2O/SiO2 molar ratio, curing temperature and age on compressive strength, morphology and microstructure of alkali-activated fly ash-based geopolymers, Cem. Concr. Compos. 33(6) (2011) 653-660. https://doi.org/10.1016/j.cemconcomp.2011.03.006

[73] M. Hu, Z. Xiaomin, F. Long, Alkali-activated fly ash-based geopolymers with zeolite or bentonite as additives, Cem. Concr. Comp. 31(10) (2009) 762-768. https://doi.org/10.1016/j.cemconcomp.2009.07.006

[74] G. Gorhan, G. Kurklu, The Influence of the NaOH solution on the properties of the fly ash-based geopolymer mortar cured at different temperatures, Composites B Eng. 58 (2014) 371-377/ https://doi.org/10.1016/j.compositesb.2013.10.082

[75] D. Hardjito, S. Fung, Fly ash-based geopolymer mortar incorporating bottom ash, Mod. Appl. Sci. 4(1) (2010) 44-52. https://doi.org/10.5539/mas.v4n1p44

[76] H. Xu, J.S.J. van Deventer, The geopolymerisation of alumino-silicate minerals, Int. J. Miner. Process. 59(9) (2000) 247-266. https://doi.org/10.1016/S0301-7516(99)00074-5

[77] S. Kramar, V. Ducman, Mechanical and microstructural characterization of geopolymer synthesized from low calcium fly ash, Chem. Ind. Chem. Eng. Q. 21 (2014) 42-44. https://doi.org/10.2298/CICEQ130725042K

[78] D. Sumajouw, D. Hardjito, S. Wallah, B. Rangan, Fly ash-based geopolymer concrete: study of slender reinforced columns, J. Mater. Sci. 42(9) (2007) 3124-3130. https://doi.org/10.1007/s10853-006-0523-8

[79] S. Ahmari, L. Zhang, Utilization of cement kiln dust (CKD) to enhance mine tailings-based geopolymer bricks, Constr. Build. Mater. 40 (2013) 1002-1011. https://doi.org/10.1016/j.conbuildmat.2012.11.069

[80] T. Sujatha, K. Kannapiran, S. Nagan, Strength assessment of heat cured geopolymer concrete slender column, Asian J. Civ. Eng. 13(5) (2012) 635-646.

[81] D. Tran, D. Kroisová, P. Louda, O. Bortnovsky, P. Bezucha, Effect of curing temperature on flexural properties of silica-based geopolymer-carbon reinforced composite, J. Achiev. Mater. Manuf. Eng. 37(2) (2009) 492-497

[82] P. Duxson, J. Provis, Designing precursors for geopolymer cements, J. Am. Ceram. Soc. 91(12) (2008) 3864-3869. https://doi.org/10.1111/j.1551-2916.2008.02787.x

[83] D. Feng, J. Provis, J.S.J. van Deventer, Thermal activation of albite for the synthesis of One-part mix geopolymers, J. Am. Ceram. Soc. 95(2) (2012) 565-572. https://doi.org/10.1111/j.1551-2916.2011.04925.x

[84] Y. Liew, H. Kamarudin, A.M. Al Bakri, M. Luqman, I.K. Nizar, C.M. Ruzaidi, C.Y. Heah, Processing and characterization of calcined kaolin cement powder, Constr. Build. Mater. 30 (2012) 794-802. https://doi.org/10.1016/j.conbuildmat.2011.12.079

[85] T. Suwan, M. Fan, Influence of OPC replacement and manufacturing procedures on the properties of self-cured geopolymer, Constr. Build. Mater. 73 (2014) 551-561. https://doi.org/10.1016/j.conbuildmat.2014.09.065

[86] Information on http://minerals.usgs.gov/minerals/pubs/commodity/cement/mcs-2009-cement.pdf

[87] P.-C. Aitcin, Cements of yesterday and today; Concrete of tomorrow, Cem. Concr. Res. 30 (2000) 1349-1359. https://doi.org/10.1016/S0008-8846(00)00365-3

[88] M. Taylor, C. Tam, D. Gielen, Energy efficiency and CO2 emissions from the global cement industry, Energy Technology Policy Division, International Energy Agency, 2006.

[89] J.S. Damtoft, J. Lukasik, D. Herfort, D. Sorrentino, E. Gartner, Sustainable development and climate change initiatives, Cem. Concr. Res. 38(2) (2008) 115-127. https://doi.org/10.1016/j.cemconres.2007.09.008

[90] J. Provis, J.S.J. van Deventer, 2014. Alkali Activated Materials. State-of-the-Art Report - RILEM TC 224-AAM. Netherlands: Springer. https://doi.org/10.1007/978-94-007-7672-2

[91] T.E. McGrath, S. Cox, M. Soutsos, D. Kong, L.P. Mee, J.U.J. Alengaram, Life cycle assessment of geopolymer concrete: A Malaysian context, IOP Conf. Ser.: Mater. Sci. Eng. 431 (2018) 092001. https://doi.org/10.1088/1757-899X/431/9/092001

[92] G. Herbert, J.B. d'Espinose de Lacaillerie, N. Roussel, An environmental evaluation of geopolymer based concrete production: reviewing current research trends, J.Cleaner Prod. 19(11) (2011) 1229-1238. https://doi.org/10.1016/j.jclepro.2011.03.012

[93] R. Bogue, The chemistry of Portland cement, Soil Sci. 79(4) (1955) 322. https://doi.org/10.1097/00010694-195504000-00014

[94] Mehta P. (1986), Concrete: Structure, Properties and Materials, New Jersey, Prentice Hall.

[95] Soroka I. (1979), Portland Cement Paste and Concrete, Old Working, Surrey, Unwin Brothers Limited. https://doi.org/10.1007/978-1-349-03994-4

[96] Mehta P., Monteiro P.J.M. (2006), Concrete: Microstructure, Properties and Materials, 3rd ed. New York, McGraw-Hill.

[97] ASTM, 2015. C125-15B Standard terminology relating to concrete and concrete aggregates. ASTM Standards Online, SUA.

[98] British Standard Institution (BSI), 2005. BS 8443:2005 Specification for establishing the suitability of special purpose concrete admixtures. British Standards Online, United Kingdom.

[99] ASRO, 2011. SR EN 197-1 Cement Part 1. Composition, specifications and conformity criteria for common cements. National Standardisation Body - ASRO, Bucharest, Romania.

[100] ASRO, 2012. SR EN 450-1 Fly ash for concrete. Definition, specifications and conformity criteria. National Standardisation Body - ASRO, Bucharest, Romania.

[101] NIRD URBAN-INCERC, 2018. PN 18 35 04 03 - Research on the valorization of mineral additions with inert, latent hydraulic or puzzolanic character in innovative cementitious compositions, in the context of the implementation of the concept of "circular economy" contributing to the creation of resilient structures in Romania, Phase 2: Design and realization of innovative cementitious compositions based on effective additions, available nationally and identifying their applicability to resilient structures. Ministry of Research and Innovation, Bucharest, Romania.

[102] CEPROCIM, 2009. C439 - The use of fly ash (type II addition) in concrete in order to improve the durability characteristics in accordance with the appropriate standards and/or regulations, harmonized with the European Construction Products Directive. Ministry of Regional Development and Tourism, Bucharest, Romania.

[103] Malhotra V.M., Mehta P. (2002), High-Performance, High-Volume Fly Ash

Concrete: Materials, Mixture Proportion, Properties, Construction Practice and Case Histories, Ed. Marquardt Printing LTD, Ottawa, Canada.

[104] Information on http://www.ecoba.com/ecobaccputil.html

[105] CEPROCIM, 2008. C50 - Study on the possibility of valorization in the Romanian economy of products resulting from the energy industry (gypsum, calcium sulphite, ash). Ministry of Economy and Finance, Bucharest, Romania.

[106] Information on http://faracarbune.ro/campania-fara-carbune/centralele-pe-carbune-din-romania/

[107] Information on http://legislatie.just.ro/Public/DetaliiDocument/51810

[108] Quattroni G., Orsenigo L., Peretti R., Zucca A., Ciccu R., Ghiani M. (1999), The Italian approach to the problem of fly ash, Ash Symposium Italy pag. 12.

[109] CEPROCIM, 1999. C39/1 - Research on the characteristics of ashes from all thermal power plants with direct influence on their behavior in cement (physical, chemical, puzzolanic activity). ANSTI, Bucharest, Romania.

[110] CEPROCIM, 2000. C39/2 - Research on the characteristics of ashes from all thermal power plants with direct influence on their behavior in cement (physical, chemical, puzzolanic activity). ANSTI, Bucharest, Romania.

[111] CEPROCIM, 2004. C255 - Solutions to optimize the puzzolanic activity of Mintia thermal power plant ash for efficient use as an addition to cement grinding. CASIAL, Deva, Romania.

[112] Lewandowski W., Feuerborn H.J. (2006), Processing plants for fly ash in Europe, ECOBA pg. 4.

[113] Willis S. (2003), Mechanically Activated Fly Ash as a Reactant in the Production of Hihg-Strength Geopolymer Mortars and Carbon-Fibre Composites, Teză de Doctorat, University of Western Australia pag.13.

[114] ASTM, 2019. C618 -Standard Specification for Coal Fly Ash and Raw or Calcined Natural Pozzolan for Use in Concrete. ASTM Standards Online, SUA.

[115] Information on https://osha.europa.eu/en/legislation/directives/regulation-ec-no-1907-2006-of-the-european-parliament-and-of-the-council

[116] ASRO, 2016. SR EN 196-1 Methods of testing cement. Determination of strength. National Standardisation Body - ASRO, Bucharest, Romania.

[117] ASRO, 2019. SR EN 12390-7 esting hardened concrete. Density of hardened concrete. National Standardisation Body - ASRO, Bucharest, Romania.

Materials Research Forum LLC
https://doi.org/10.21741/9781644901533

[118] Black J.R. (2012), Mix Design Process for Alkaline-Activated Class F Fly Ash Geopolymer Concrete, Final Project Report, University of New South Wales, Australian Defence Force Academy.

[119] S.R. Ahmari, X. Toufigh, L. Zhang, Production of Geopolymer Binder from Blended Waste Concrete Powder and Fly Ash, Constr. Build. Mater. 35 (2012) 718-729. https://doi.org/10.1016/j.conbuildmat.2012.04.044

[120] J. Tailby, K. MacKenzie, Structure and Mechanical Properties of Aluminosilicate Geopolymeric Composites with Portland Cement and its Constituent Minerals, Cem. Conc. Res. 40 (2010) 787-794. https://doi.org/10.1016/j.cemconres.2009.12.003

[121] M. Criado, A. Fernandez-Jimenez, A. Palomo, Alkali-activation of fly-ash: effect of the SiO2/Na2O ratio, Micropor. Mesopor. Mat. 106 (2007) 180-191. https://doi.org/10.1016/j.micromeso.2007.02.055

[122] M. Criado, A. Fernandez-Jimenez, A. Palomo, (2010), Alkali-activation of fly ash: Part III: Effect of curing conditions on reaction and its graphical description, Fuel 89 (2010) 3185-3192. https://doi.org/10.1016/j.fuel.2010.03.051

[123] M. Olivia, P.K. Sarker, H. Nikraz, Water Pentrability of Low Calcium Fly Ash Geopolymer Concrete, Proceedings of the International Conference on Construction and Building Technology, Kuala Lumpur, Malaysia (2008) 517-530.

[124] J. Wongpa, K. Kiattikomol, C. Jaturapitakkul, P. Chindaprasirt, Compressive strength, modulus of elasticity and water permeability of inorganic polymer concrete, Mater. Des. 31 (2010) 4748-4754. https://doi.org/10.1016/j.matdes.2010.05.012

[125] C. Shi, P.V. Krivenko, D.M. Roy, 2006. Alkali-Activated Cements and Concretes. Abington, UK: Taylor & Francis. https://doi.org/10.4324/9780203390672

[126] P. Krivenko,1997. Alkaline cements: terminology, classification, aspects of durability. Proceedings of the 10th International Congress on the Chemistry of Cement, Gothenburg, Sweden.

[127] H. Xu, J.L. Provis, J.S.J. van Deventer, P.V. Krivenko, Characterization of aged slag concretes, ACI Mater. J. 105(2) (2008) 131-139. https://doi.org/10.14359/19753

[128] D. Hardjito, S.E. Wallah, D.M.J. Sumajouw, B.V. Rangan, Fly ash-based geopolymer concrete, Aust. J. Struct. Eng. 6 (2005) 77-86. https://doi.org/10.1080/13287982.2005.11464946

[129] T. Stengel, J. Reger, D. Heinz, Life cycle assessment of geopolymer concrete - What is the environmental benefit?, Conc. Sol. 09 (20019).

[130] ASRO, 2004. SR EN 1338 Concrete paving blocks. Requirements and test methods. National Standardisation Body - ASRO, Bucharest, Romania.

[131] Information on https://isc.gov.ro/files/2017/Piata/Hotarare%20668%202017.pdf

[132] Information on https://eur-lex.europa.eu/legal-content/ro/ALL/?uri=CELEX:32011R0305

List of Symbols

CCP	-	COAL COMBUSTION PRODUCTS;
FA	-	Fly ash;
BA	-	Furnace bottom ash;
BS	-	Boiler slag;
FBC	-	Fluidized bed combustion ash;
SDA	-	Semi dry absorption product;
FDG	-	Flue gas desulphuration gypsum;
L.O.I.	-	Loss on Ignition;
NaOH	-	Sodium hydroxide;
Na_2SiO_3	-	Sodium silicate;
AAGM	-	Alkali-activated geopolymer material;
$R_{0,045}$	-	Fineness on the 0,045 mm sieve;
AAGP	-	Alkali-activated fly ash-based geopolymer paste;
AAGC	-	Alkali-activated fly ash-based geopolymer concrete;
AA	-	Alkaline activator;
SH	-	Sodium hydroxide solution;
SS	-	Sodium silicate solution;
D	-	Apparent density, $kg/m3$;
m	-	Mass of the sample, g;
V	-	Volume of the sample, m^3;
$R_{ti}med$	-	3PB flexural strength, N/mm^2;
P_{med}	-	Bending break force, kN;
L	-	Dimensions of the samples, mm;
b	-	Dimensions of the samples, mm;
l	-	Distance between lower rollers, mm;
$R_c med$	-	Compressive strength of the samples, N/mm^2;

F_{med}	-	Compression maximum force, kN;
A	-	Square section area of the loading device, mm^2;
AA/FA	-	Alkaline activator to fly ash ratio;
SS/SH	-	Na_2SiO_3 to NaOH solution ratio;
M_1	-	Mass in saturated state, g;
M_2	-	Mass in dry state, g;
W_a	-	Total water absorbtion, %;
L	-	Loss in mass per unit area (freeze-thaw), kg/m^2
M	-	Mass of the total quantity of peeled material after 28 cycles, kg;
A	-	Surface area of the sample, m^2
ΔV	-	Loss in volume after 16 cycles (Böhme abrasion test), mm^3;
Δm	-	Loss in mass after 16 cycles, g;
$ρ_R$	-	Density of the sample, g/mm^3.
T	-	Tensile splitting strength, MPa;
P	-	Maximum breaking load, N;
K	-	Correction factor for the thickness of pavers;
S	-	Loading surface area, mm^2;
L	-	Average of two length measurements, one on the top and one on the bottom of the paving block, mm;
T	-	Thickness of the paving block, mm;
F	-	Load per length unit, N/mm;

About the Authors

PhD Eng Adrian Lăzărescu is Scientific Researcher at N.I.R.D. URBAN-INCERC, Cluj-Napoca Branch, Romania and PhD in Civil Engineering, granted at Technical University of Cluj-Napoca in the field of alkali activated geopolymer materials. His main interest is to study the possibility of producing this type of material using Romanian local materials, with the aim to assess the possibility of using fly ash as raw material in the production of innovative, new-alternative materials in the civil engineering field. Together with the team co-authors, the author was a member in several National Projects which targeted the use of industrial by-products in the production of alternative materials, within the Circular Economy and Sustainable Development principles. Also, within the research team of N.I.R.D. URBAN-INCERC, the author participated in extensive research programs on the development of cementitious materials with self-healing properties of microcracks (ECC – Engineered Cementitious Composites with Self-Healing Properties), self-healing cementitious materials, alternative thermal insulation products based on sheep wool and has coordinated numerous in-situ NDT and destructive investigations of buildings, respectively research on the quality of materials and products used in the construction industry.

PhD Eng Henriette Szilagyi is Senior Researcher and also the Manager of N.I.R.D. URBAN-INCERC, Cluj-Napoca Branch, Romania. A renowned practitioner in the field of concrete technology, her main research involves special concrete (self-compacting concrete, ultra-high-performance concrete, lightweight concrete, concrete with waste and by-products such as recycled glass, fly ash, recycled plastic and other innovative cementitious materials) and mortars, cements, admixtures, additives, building materials. She has published extensively on many aspects regarding concrete technology and sustainable development in the construction industry. She is Associate professor at Technical University Cluj-Napoca, Faculty of Civil Engineering; she also mentoring students in research projects, guidance for MSc Dissertation, PhD Thesis.

PhD. Eng Cornelia Baeră is Scientific Researcher at N.I.R.D. URBAN-INCERC, Timișoara Branch, Romania. Her main research interests involve the topic of high-performance concrete and mortars (Fibre reinforced concrete, Concrete with self-healing properties, Concrete with mineral additions, green concrete, Engineered cementitious composites (ECCs), etc.). The doctoral studies were performed at T.U. Cluj-Napoca, when she worked as a researcher at INCD URBAN-INCERC Cluj-Napoca Branch together with the team co-authors. Currently, her research interests extended towards the topic of Management in Production and Transportation, at Research Center in Engineering and Management, Politehnica University of Timișoara.

PhD Eng Andreea Hegyi is Scientific Researcher at N.I.R.D. URBAN-INCERC Cluj-Napoca Branch, Romania. Her main research interests involve studies regarding the durability of cementitious composites using hot-dip galvanized rebars, the development of ecological materials for the construction industry, studies regarding the parametric evaluation of self-cleaning cementitious composites using nano-TiO_2 particles and the development of alternative thermal insulation materials based on sheep wool. Her expertise in research on the quality of materials and products used in the construction industry has opened new perspectives in studying new and alternative construction materials which comply with current legislation and further improving of existing ones in the field by addopting alternative solutions and by creating a "bridge" between traditional and modern construction materials.

www.ingramcontent.com/pod-product-compliance
Lightning Source LLC
Chambersburg PA
CBHW071704210326
41597CB00017B/2327